DATA QUALITY PROBLEMS IN ARMY LOGISTICS
Classification, Examples, and Solutions

Lionel A. Galway

Christopher H. Hanks

Prepared for the United States Army

Arroyo Center

RAND

Approved for public release; distribution unlimited

UC
263
.G35
1996

PREFACE

This is the final report of the project "Logistics Information Requirements and Quality." The research was jointly sponsored by the Logistics Support Activity within the U.S. Army Materiel Command and the Director for Plans and Operations, Office of the Deputy Chief of Staff for Logistics within the Army Staff. The research was conducted in the Arroyo Center's Military Logistics Program. The Arroyo Center is a federally funded research and development center sponsored by the United States Army. The report should be of interest to logisticians at all levels of the Army and the DoD, particularly those who are responsible for the operation and development of logistics information systems. More broadly, it should also interest people who have responsibilities for data quality in large, complex organizations.

CONTENTS

Preface iii
Figures vii
Tables ix
Summary xi
Acknowledgments xv
Abbreviations xvii

Chapter One
    DATA QUALITY PROBLEMS IN ARMY LOGISTICS 1
    Introduction 1
    Data Versus Information 2
    Importance of Data Quality 3
    Army Initiatives and Data Quality 3
    Commercial Organizations 4
    Organization of the Report 6

Chapter Two
    METHODOLOGY 7
    Starting Point 7
    Data Elements Versus Databases 7
    Selected Data Elements 8
    Data Tracking 9

Chapter Three
    CLASSIFYING DATA PROBLEMS 11
    Definition of Data Quality 11
    Classification of Data Problems 12

Operational Data Problems	12
Conceptual Problems	13
Organizational Data Problems	15

Chapter Four
EXAMPLE OF LOGISTICS DATA QUALITY PROBLEMS:
THE END ITEM CODE	17
Introduction	17
Description of the EIC	18
History and Purpose of the EIC	18
Structure and Assignment of the EIC	19
EIC Data Flow in Supply and Maintenance	20
Current Uses of the EIC/CDDB	23
Operational Problems, Causes, and Attempted Fixes	26
Early Problems	26
EIC Enhancement	27
Maintenance Master Data File	30
An Unresolved Problem: PLL Replacements	31
Conceptual Problems, Causes, and Attempted Fixes	33
Solutions for Operational and Conceptual Problems	38
Operational Solutions	38
Conceptual Solutions	39

Chapter Five
ORGANIZATIONAL DATA PROBLEMS	41
Evidence of Organizational Problems	41
Hypotheses About Causes of Organizational Data Problems	42
Evidence for the Hypotheses	43
Effects on Data Quality	45
Two Proposals for Fixing Organizational Problems	48

Chapter Six
CONCLUSIONS AND RECOMMENDATIONS	51
Operational and Conceptual Problems	51
Organizational Problems	52
Data Quality and Information Quality: Implications for Force XXI and Velocity Management	54

Appendix: OTHER DATA ELEMENTS	57
References	67

FIGURES

4.1. EIC Data Flow to CDDB 21
4.2. PLL Replacement Requisitions Do Not Carry an EIC . . 32

TABLES

3.1.	Typology of Data-Quality Problems	12
4.1.	EIC Missing Rates/Enhancement Effect in Sample CDDB Data	29
A.1.	Valid, Missing, and Invalid Failure Codes	59
A.2.	Breakdown of Valid Failure Codes	60
A.3.	Percentage of Missing Failure Codes by MSB Company	62

SUMMARY

— information failure —

Implicit in the Army Force XXI concept is the idea that information and data are assets—as important in their own right as the Army's physical assets of personnel, vehicles, and materiel. The Army's Velocity Management initiative, which is focused on logistics, recognizes the central importance of using performance data to inform the reengineering and management of logistics processes. To truly qualify as an asset, however, data must have the requisite quality.

Unfortunately, much logistics data in the Army is widely perceived to be of poor quality. This perception is based on personal experience, anecdotes, and numerous examples of failed analyses and modeling efforts that were unable to overcome data problems. The purpose of this project was to examine quality problems in Army logistics data and to recommend solutions. To focus the project, we selected a small group of data elements generated by the "retail" Army that are transmitted to and used by centralized logistics activities in the "wholesale" Army.

Our working definition of "bad data" is based on the current data-quality literature, which links the idea of data quality to the uses to which data are put: if a given set of reported data cannot provide the information needed for decisions, a data-quality problem exists. Our discussion of logistics data problems is grounded in the uses of the data.

Data-quality problems should be properly classified so they can be solved. We propose a three-level framework for understanding and classifying the nature of data problems:

- *Operational* data problems are present when data values are missing, invalid, or inaccurate.

- *Conceptual* data problems are present when the data, because of imprecision or ambiguities in their definition, are not suitable for an intended use or, because of definitional problems, have been subjected to varying collection practices, again resulting in missing, invalid, inaccurate, or unreliable values.

- *Organizational* data problems occur when there are disconnects between the various organizations that generate and use data, resulting in a lack of agreement on how to define and maintain data quality. One symptom of organizational problems is the persistence of operational and conceptual problems over time, even after repeated attempts at solution.

We describe in detail the quality problems with the End Item Code (EIC). (Other logistics data elements are treated more briefly in an appendix). The EIC is a three-character code that serves as the key data element in the Army's Central Demand Data Base (CDDB). The central purpose of the CDDB is to record the link between parts and end items when a part is demanded to fix a broken end item. The EIC is used in maintenance data systems as well to tie repair actions to end items. The EIC has data problems of all three types.

Operational. The EIC has a fairly high blank (missing value) rate in the CDDB (currently upwards of 50 percent). This is in spite of several attempts to fix operational problems, including official exhortation, an "EIC enhancement" program run at the wholesale level using ancillary data files (which has revealed that many nonblank reported EICs are incorrect as well), and a new initiative that enters the EIC automatically (for selected end items) on individual requests for issue. However, in the course of a visit to the retail level (Fort Riley) we learned that one-for-one unit Prescribed Load List (PLL) replacement demands (which occur every time the PLL successfully issues a part) do *not* carry the associated EIC forward for transmission to the CDDB. This would account for a substantial proportion of blank EICs in retail-level demands. While this problem has since been fixed, it appears to have been unknown and unsuspected by the wholesale system since 1987, when the EIC and the CDDB were created.

Conceptual. The EIC is used to relate parts demands and repair actions to end items. "End items" in the Army are equipment items subject to the procurement process at Army Materiel Command Major Subordinate Commands (MSCs). (As a result, there are about 8,700 end items with distinct EICs). End items are not the same thing as "weapon systems." This causes conceptual problems when attempts are made to use the EIC to obtain weapon-system-level information: new users obtain results that are often incomplete or wrong, while more experienced users give up trying to use EIC data to get weapon-system information.

Organizational. The persistence of missing, invalid, and inaccurate EIC values in the CDDB, and the fact that weapon-system information has never been captured by the code, lead us to suggest that the EIC (and other problem data elements collected by the retail level for use by the wholesale system) have deeper problems: we hypothesize two related causes of organizational problems. First, the gap between the wholesale Army and the retail Army is so wide and deep that agreement between the two on data-quality issues is difficult. This leads to a lack of visibility of data benefits by the retail side, and a lack of visibility of data costs (data burden) by the wholesale side. Second, the important negotiation that should occur between wholesale and the retail level must take place in the complex organizational context of logistics information system development, involving FORSCOM, TRADOC, AMC, the Information Systems Command, and the Army acquisition infrastructure. This fragmentation blurs data issues by confounding them with the separate issues of hardware and software development.

We suggest that the organizational problem could be resolved (leading to more lasting and complete solutions of the operational and conceptual problems) by negotiating data issues directly between the retail level and wholesale system. This may require new approaches, such as having the wholesale system pay the retail level to provide high-quality data.

ACKNOWLEDGMENTS

In the course of this project we interviewed many people at all levels of the Army, all of whom were extremely cordial and spent a great deal of time to help us understand the impact of logistics data quality on their work. Below we have listed as many of them as we can. To a large extent, they have made possible the large-scale picture we have put together of the Army's logistics data problems. Any errors in interpretation and the suggestions for improvement are strictly the responsibility of the authors.

- **LOGSA.** Dan Rhodes, Dan McMillan, Marlene Ellul, COL Norman Myers, Skip Popp, Paulette Myers, Marcia Byrnes, Robert Stauner, Steve Tesh, Byron Sickler, Tom Ress, Dennis Blenman, Michael Lainhart, Eric Hinson, Ken Walker, Connie Lainhart.
- **DCSLOG.** Mike Blackman.
- **Fort Riley, KS.** At the 701st MSB: CW3 Jeff Martin, CPT Chris Livingston, LTC Vincent Boles, LTC Pat McQuistion, SGT Keith Asher, SPC Chuck Higman, CW2 Robin Pitts, Sgt. Shields, Sgt Rawls. At the 101st FSB: CW1 Durham.
- **CASCOM.** COL Mike Topp, Brian Woods.
- **PM-ULLS.** Nick Flain.
- **DCL.** MAJ Doug Mosher, Jack Smith, Gerard Rodriguez, Dot Walton, Bill Dooley.
- **MICOM.** Tom Ingram, Norbert Lutz, Alfie Onuszkanycz, SFC Gary Waggoner, Bill Facundo.

- **TACOM.** Tony Cuneo.
- **ATCOM.** Teddie Stokes, Bill MacDonald, Roger Hoffman, Cheryl Hammond, Pat Lawson, Dan Kruvald, Gerald Koepke, Norm Reinagel, Thom Blakely, Jon Evans.
- **AMSAA.** Clarke Fox, Maria Zmurkewycz, Jeff Landis, Greg Serabo, Vickie Evering, Jin Kwon, David Chung, Ed Gotwals, Ruth Dumer.

Many other people we talked to briefly on the phone, tracking down different aspects of data use across the Army and in various parts of the government.

We also thank the following individuals: Brandt Adams, Ed Beyer, and Don Taylor of CALIBRE Systems, Inc., Charles Davis and Carl Hayes of DATA, Inc., our RAND colleagues Marygail Brauner, John Folkeson, and Marc Robbins, and the participants in the Payday Logistics Seminar. All made valuable comments on our research and draft versions of this report.

We also thank Kip Miller of RAND and Thomas Redman (formerly of AT&T) for formal reviews of the report and many helpful comments.

ABBREVIATIONS

AMC	Army Materiel Command
AMCCOM	Army Munitions and Chemical Command
AMDF	Army Master Data File
AMSAA	Army Materiel Systems Analysis Activity
AOAP	Army Oil Analysis Program
ASL	Authorized Stockage List
ATCOM	Aviation and Troop Command
CASCOM	Combined Arms Support Command
CCSS	Commodity Command Standard System
CDDB	Central Demand Data Base
CECOM	Communications and Electronics Command
DAAS	Defense Automated Addressing System
DBOF	Defense Business Operations Fund
DCL	Development Center Lee
DCSLOG	Deputy Chief of Staff for Logistics
DODAAC	Department of Defense Activity Address Code
DS	Direct Support
EDF	Equipment Data File
EIC	End Item Code

ERC	Equipment Readiness Code
FORSCOM	Forces Command
FSB	Forward Support Battalion
GAO	Government Accounting Office
GS	General Support
HEMTTS	Heavy Expanded Mobility Tactical Truck
HMMWV	High Mobility Multi-Wheeled Vehicle
LCA	Logistics Control Agency
LEA	Logistics Evaluation Agency
LIF	Logistics Intelligence File
LIN	Line Item Number
LMI	Logistics Management Institute
LOGSA	Logistics Support Activity
LRU	Line Replaceable Unit
MICOM	Missile Command
MMDF	Maintenance Master Data File
MOS	Military Occupational Specialty
MRSA	Materiel Readiness Support Activity
MSB	Main Support Battalion
MSC	Major Subordinate Commands
NHA	Next Higher Assembly
NICP	National Inventory Control Point
NIIN	National Item Identification Number
NSN	National Stock Number
OPTEMPO	Operational Tempo
OSD	Office of the Secretary of Defense
OSMIS	Operating and Support Management Information System

OSRAP	Optimum Stockage Requirements Analysis Program
PLL	Prescribed Load List
PMR	Provisioning Master Record
SAILS	Standard Army Intermediate Logistics System
SAMS	Standard Army Maintenance System
SARSS	Standard Army Retail Supply System
SDC	Sample Data Collection
SESAME	Selected Essential Item Stockage for Availability Method
SIMA	Systems Integration and Management Activity
SIWSM	Secondary Item Weapon System Management
SLAC	Supply List Allowance Computation
SMS	Standard Maintenance System
SOP	Standard Operating Procedure
SORTS	Status of Resources and Training System
TACOM	Tank and Automotive Command
TRADOC	Training and Doctrine Command
TUFMIS	Tactical Unit Financial Management Information System
UIC	Unit Identification Code
ULLS	Unit Level Logistics System
USACEAC	U.S. Army Cost and Economic Analysis Center
VAMOSC	Visibility and Management of Operating and Support Costs
WOLF	Work Order Logistics File
WRAP	War Reserve Automated Process
WS/EDC	Weapon System/Equipment Designator Code

Chapter One
DATA QUALITY PROBLEMS IN ARMY LOGISTICS

INTRODUCTION

This report addresses problems in the quality of Army logistics data and information (we will make the distinction between these two shortly), the causes of those problems, and potential fixes. It is widely perceived in the Army that severe problems exist with the logistics data that provide the basis for many important Army decisions. Field commanders at all levels complain about a lack of visibility of requisition and shipment status. The wholesale system complains about missing data in requisitions and maintenance reports, which makes it difficult to get a broad picture of how Army equipment is performing around the world. Efforts to build computerized decision support systems to aid in logistics decisionmaking have foundered on inadequate data.

Data are largely intangible: jottings on paper, electronic entries in a database, characters on a video screen, recollections from a person's memory. Particularly in the Army, data seem insignificant compared to the physical assets of equipment, personnel, and materiel. However, data are also assets: they have real value when they are used to support critical decisions, and they also cost real money to collect, store, and transmit. The quality of the data the Army uses to manage its logistics processes has real impacts on how well that management is done.

Data and information are of special concern because their extensive use is one of the keys to success of major new Army initiatives such as Force XXI and Velocity Management (VM).[1] With these initiatives, the Army will depend on highly accurate information communicated to all levels of command to control a force with potentially widely dispersed operational and support forces. For this "force digitization" to work, the data and information which form its foundation must be of high enough quality to support the decisions being made.

DATA VERSUS INFORMATION

We make the following distinction between data and information: data, or data elements, are specific entries in a database or an information system (usually electronic, but also paper-based); information is the combining of different pieces of data to produce new quantities that provide insight into the processes producing the data. For example, maintenance systems may record data such as the start and end dates of a series of maintenance actions on a particular component. Computing the mean of the elapsed times over many repairs produces information, showing how long it takes to repair the component on average over many different attempts. Information may become data again, if it is recorded and used for further analysis, so the transition from data to information is not always one-way or necessarily a single-step process.

We are ultimately interested in the quality of information, but the quality of the underlying data is clearly crucial to the quality of the derived information. Looking at data elements has the advantage of providing a firm anchor point in the data-to-information flow: by focusing on a data element and its quality, we can ask what information that data element is used to generate and how the quality of that information depends on the quality of the data. This viewpoint is particularly important when several different uses are made of the same data element and it is necessary to assess the effect of its quality in different contexts. Therefore, in this study we organize our discussion around data elements and their quality problems, but we do

---

[1]Evaluation of the recent Focus Dispatch exercise indicated that CS digitization was one of the biggest successes of the effort (Naylor, 1995).

so in the context of the use of those data elements in producing information.

IMPORTANCE OF DATA QUALITY

It is easy to think of important logistics decisions that must be made based on data collected from maintenance, supply, and transportation organizations. For example, detecting an increase in the numbers of particular failures can alert engineers to the potential need for a modification, and analysis of the specific circumstances (i.e., what failed and under what conditions) can help specify the modification. Analysis of transportation times can help pinpoint bottlenecks. Against these benefits, however, logistics data also has costs: it takes time to capture data, money to buy the technology to do the capture, and people and money to staff and support the systems that store data, transmit it, and analyze it. Thus, while good data are needed for good decisions, the cost of getting good data must be weighed against the benefits. This tradeoff between quality and costs can be complex.

The costs to collect data and to ensure quality (e.g., detailed edit checks at the point of entry) are often very visible to the collecting entity in terms of time and energy expended. The benefits may be very diffuse, however, particularly in a large organization like the Army, where data collected in one place may be analyzed and used in very distant parts of the organization with different responsibilities and perspectives. In these cases, one part of the organization may be asked or required to collect data that have little immediate effect on its own operations but that can be used by other parts of the organization to make decisions with long-term impacts. Intraorganizational incentives and feedback to insure data quality in these cases have been difficult to devise.[2]

ARMY INITIATIVES AND DATA QUALITY

The issue of data quality, particularly for logistics data, is an important one for two current Army initiatives: Force XXI and Velocity

---

[2]Redman (1995).

Management. Both are attempting to radically change the Army, and both depend heavily on information and data technology to do so.

The digitization initiatives in Force XXI aim to radically upgrade communications and information-processing technology in all parts of the Army and to integrate them into all areas of operations and support. The explicit expectation is that by upgrading the availability and quantity of data, forces will become more efficient and effective by improving processes while reducing required personnel and materiel resources. Information (and the underlying data) is thus explicitly assumed to be an asset that can be substituted for other, more expensive assets.[3] But this requires that the information and data be of the requisite quality. In logistics, our research indicates that meeting these expectations will require a great deal of work to improve the current poor quality of logistics data.

Velocity Management[4] is explicitly focused on logistics; its goal is to reengineer and improve support functions by establishing baselines, identifying sources of inefficiencies, setting goals for corrective actions to be taken, and measuring performance. Each of these tasks requires good data. Obtaining good baseline data from current Army data systems has proved to be a problem; monitoring and evaluating improved processes may require rethinking and reengineering logistics data systems as well.[5]

COMMERCIAL ORGANIZATIONS

The literature on evaluating and improving data quality is relatively new, dating back only to the mid-1970s with work done for the Energy Information Administration (EIA) on the quality of a set of surveys of the nation's energy resources.[6] The topic of data quality

---

[3]Although the emphasis in Force XXI documentation is on operational information (see, e.g., TRADOC PAM 525-5), logistics information can also substitute for physical assets. "Total asset visibility," for example, is expected to reduce stockage requirements.

[4]Dumond, Eden, and Folkeson (1995).

[5]The other services also have logistics data-quality problems to deal with. See Abell and Finnegan (1993).

[6]See Energy Information Administration (1983) for an overview of the entire set of reviews.

reemerged as an important issue in commercial and government organizations in the late 1980s. Commercial organizations were driven by strong competitive pressures to reengineer and improve their business processes, and data began to be seen as a key asset in this drive.[7] The explosive increase in computer networking had given people access to a much wider array of databases, leading to an increased awareness that much of the available data was of questionable if not poor quality. This has been highlighted by data-quality studies of scientific, medical, justice, and business databases and by the occurrence of some very expensive business mistakes.[8] Prominent contributions to the academic literature in the field have come from the work of Redman and his colleagues at AT&T,[9] and from the program in Total Data Quality Management at MIT, directed by Wang.[10]

The consensus of the data-quality research is that while there are general approaches that cut across all areas (e.g., the creation and maintenance of *metadata* that describe the origin and quality of data in a database[11]), most data-quality problems are best addressed in the context of the particular processes that generate and use the data. Thus, for example, although the AT&T Quality Steering Committee has supported research aimed at characterizing data-quality problems in the abstract,[12] its main focus has been to formulate a set of guidelines for improving data quality to be used by process action teams within the process itself.[13]

---

[7] Redman (1995).

[8] A manufacturer found that salesmen using a new software system had created new account numbers for each sale made, splitting the records of large customers across hundreds or thousands of records (Bulkeley, 1992). See also Blazek (1993), Laudon (1986), Kolata (1994), and Hardjono (1993).

[9] Redman (1992), AT&T Quality Steering Committee (1992a, 1992b, 1992c).

[10] Hansen and Wang (1991), Wang and Kon (1992).

[11] For example, Wang, Kon, and Madnick (1992).

[12] Fox, Levitin, and Redman (1994), Levitin and Redman (1995), and Levitin (undated).

[13] AT&T Quality Steering Committee (1992a, 1992b, 1992c).

ORGANIZATION OF THE REPORT

In this first chapter we have discussed why data quality is important to the Army. Chapter Two describes the methodology of the study. Chapter Three defines data quality more precisely and, based on our research findings, outlines a three-level framework for classifying data problems. We use that framework in Chapter Four to organize the discussion of an important representative data element, the End Item Code, which exhibits all three types of problems. Chapter Five discusses the deeper issue of how to fix the most difficult problems, which we argue are more important, more subtle to detect, and harder to solve. Chapter Six summarizes our conclusions and suggests areas for future research. An appendix contains findings on data-quality problems, causes, and fixes for several other data elements in addition to the EIC.

Chapter Two
METHODOLOGY

STARTING POINT

The Army logistics system is complex, with a large number of decisionmakers and information systems. Since data-quality problems are particularly acute when data are collected in one organization for use by another, we focused on decisionmaking at the wholesale level using data supplied by the retail level. Our starting point was the Logistics Support Activity (LOGSA), a field activity of the Army Materiel Command. LOGSA maintains databases containing most of the retail logistics data of interest to the wholesale system, including the Central Demand Data Base (CDDB) in supply, the Work Order Logistics File (WOLF) in maintenance, and the Logistics Intelligence File (LIF) in supply and distribution. We chose LOGSA as a strategic storage point in the data flow between retail and wholesale, where we could access the data, identify wholesale users, and take advantage of knowledge at LOGSA about data-quality problems in the databases.

DATA ELEMENTS VERSUS DATABASES

Early in the project we faced the decision of whether to exhaustively analyze one specific database such as the WOLF, or to look at multiple problem data elements across many databases. The argument for focusing on one data system extended the storage point argument: by working with a single database at LOGSA, we could start the project with a tight focus and expand as needed. LOGSA, however, was embarked on an effort to reorganize and streamline its databases, so the future existence of specific individual databases was uncertain. Further, by looking at only one database, we ran the risk of expending effort on redundant or unused data elements that might well be eliminated in the course of LOGSA's database review.

Based on discussions with our sponsors at LOGSA and ODCSLOG, we decided as an alternative to focus on a set of selected data elements from several databases. Our criteria were that the data elements should

- be perceived as having significant quality problems
- be currently used for decisions by the wholesale system, or at least be potentially useful (if quality problems could be resolved)
- be representative of broader data-quality problems.

After overview briefings from LOGSA on their most important databases and data elements, and discussions of problems that both LOGSA and its customers had encountered with particular data elements, we selected a small set on which to focus our study.

SELECTED DATA ELEMENTS

Of all the data elements we discussed, the *End Item Code* (EIC) was the most important. It played a central role in the Central Demand Data Base (CDDB) and was also important for certain types of analyses that were performed using data from WOLF. It had a long history of problems, and much attention had been devoted to fixing to those problems, with limited success. For all of these reasons, this data element became the central focus of our work.

There was a second set of data elements that were also perceived to have fairly serious data problems, but which were not as critical or visible as the EIC because their uses were more specialized. These were *failure codes* in maintenance systems, the *Military Occupational Specialty (MOS) accomplishing repair*, the *list of parts used* in repair actions, and *serial number* information on items undergoing repair. In some cases, such as the failure code, the quality problems were so severe as to make it virtually unusable. We examined the failure code and MOS problems in some detail, but we found that for the other two, quantitatively assessing quality proved to be so complicated that it was difficult to define, let alone assess, data quality.

Finally, there was a third tier of data elements that LOGSA cited as needing attention for data quality, but which either involved complex definitional issues or were, on closer examination, an entire class of related data problems. The *LRU indicator,* for example, has a critical use in deciding the contents of contingency packages because it indicates what repair parts can be replaced by units. However, its definition is related to several other material codes. *Usage data,* the amount of activity experienced by a vehicle or weapon system, is not measured well currently, but its collection is undergoing significant change (and depends on the EIC). *Timeliness* is a data-quality problem, but its causes and effects depend on the data element that is untimely. Finally, LOGSA was interested in the potential of collecting *organizational-level maintenance data;* our work with failure codes indicates some of the quality problems this would pose, but we did not address this topic in any further detail.

DATA TRACKING

Our method for studying data-quality problems is based on a "data tracking" method proposed by Redman and his colleagues at AT&T Bell Laboratories.[1] In this method, a data element is tracked from creation to final use, and problems and inaccuracies are noted as they are introduced and discovered along the way. Attention focuses, however, not only on the entry and manipulations performed on a data element as it is transmitted, stored, and used, but also on the organizations it transits, how the organizations view the data, and their relationships with the other organizations in the data flow path.

Since we were working primarily with data elements in the CDDB and WOLF, we started with discussions at LOGSA about the structure of the databases and their analyses of the problems with the specific data elements. We used LOGSA records of users who had requested data including these elements to conduct interviews by phone and in person with users at Army Materiel Command's (AMC) Major Sub-

---

[1] Redman (1992, 1994).

ordinate Commands: the Tank and Automotive Command (TACOM), the Aviation and Troop Command (ATCOM), and the Missile Command (MICOM), as well as the National Guard Bureau, the Government Accounting Office (GAO), various retail-level users, and contractors such as CALIBRE Systems, Inc. We asked all of these users to describe how the data elements had or had not met their requirements. We also visited the Combined Arms Support Command (CASCOM) and Development Center Lee (DCL), the organizations responsible for specifying and implementing the retail-level logistics information systems (Unit Level Logistics System and Standard Army Maintenance System (ULLS and SAMS)) where the data elements are captured prior to transmission to LOGSA. We also conducted a three-day visit to Fort Riley, Kansas, where we spent time with ULLS and SAMS supervisors and clerks who walked us through the data capture and entry procedures for the elements we were interested in. Finally, we analyzed extracts of SAMS and CDDB data and conducted telephone interviews with the units represented in the data so that we could understand the patterns of problems we observed.

Chapter Three

CLASSIFYING DATA PROBLEMS

DEFINITION OF DATA QUALITY

It is easy to elicit anecdotes about poor data and their effects, but much harder to come up with a general yet precise definition of what it means for data to be "bad." One line of academic research has attempted to determine the attributes of "good" data.[1] Another has looked at various aspects of data and evaluated how those aspects affect quality.[2] While these studies have attracted some interest, most researchers have settled on a pragmatic, usage-based definition of data quality. In this view, which we will adopt in this report, data quality can only be evaluated in the context of a use or set of uses. (It follows that data appropriate for one use may not be appropriate for another. One of the primary reasons why data-quality problems occur is that data are used for purposes not intended or envisioned when they were designed or collected.) Although we will discuss the accuracy, timeliness, definition, consistency, etc. of individual data elements, the starting point will always be a set of current or planned uses, and how the data element, as currently defined, collected or aggregated, cannot meet the requirements of that use.

---

[1] Wang and Guarascio (1991).

[2] Levitin (undated). See Redman (1992) for a comprehensive treatment of data quality from this perspective.

CLASSIFICATION OF DATA PROBLEMS

While the quality of specific data elements may be determined by the set of uses to which the data are put, it is possible to generalize about the *types* of data-quality problems encountered. Based on our study of data-quality problems in Army logistics data, we have developed a typology of problems based on their causes and symptoms.[3] This typology is laid out in Table 3.1.

We discuss each of these types of problems in more detail below, and use them in Chapters Four and Five and the appendix to evaluate data-quality problems with specific data elements in Army logistics.

Operational Data Problems

A data element has operational problems if it is missing, incorrect, invalid, or inaccurate to such an extent that it cannot be used for making the decisions under consideration. This includes most of the situations usually considered to be bad data.[4] There is an implied presumption that, were the data correct, the user could directly utilize them with no further *data* problems in making the necessary decision(s). Also implicit is the idea that there is a "correct" value

Table 3.1

Typology of Data-Quality Problems

Type	Symptoms	Causes
Operational	Data are missing, inaccurate, or invalid	Problems with data capture or transmission
Conceptual	Data are missing, inaccurate, or invalid	Data not well defined or not suitable for intended use
Organizational	Persistent operational or conceptual problems	Disconnects between organizations that collect and use the data

---

[3]We regard this typology as only a first step in classifying data-quality problems; it will be refined in future research. However, based on our work here, even this rough typology provides important insights into the causes of data problems and helps to identify what methods are useful in fixing them.

[4]Much of the AT&T material for workers focuses on operational problems.

that is measurable at least in theory (although possibly difficult or expensive to actually measure in practice with the required accuracy).

Pure operational data problems are the easiest to fix in principle: modify the method of data collection to capture the correct data. This might mean using new technology for collection (e.g., using a bar code reader rather than having a person keypunch or write down a number), entering a data element only once and deriving further instances from that entry, performing edit and consistency checks at the entry point when errors can be easily corrected, or improving coding schemes to reduce ambiguity (e.g., by adding a code for "not applicable" to distinguish between situations in which a code cannot be assigned from those in which it has been omitted). Operational errors can sometimes be corrected by downstream quality checks, where data from other sources can be used to check consistency, although fixing operational problems is best done at the source.[5]

Conceptual Problems

Data have conceptual problems when they are not well defined or are not suitable for their intended or putative use (even when completely purged of operational problems). Examples are data elements where the definition of what is being measured is imprecise, where the end user does not understand critical aspects of the data collection process, or where coding is done based on local interpretations that may vary in unknown ways from site to site.

For example, the EIA-sponsored data validation surveys have noted numerous instances of vague and hard-to-operationalize concepts in energy surveys. One such concept is that of the energy imports into and out of a particular state. This is an easy idea to state, but very difficult to measure for states that share a large metropolitan area with another state.[6] In these cities, large energy suppliers may be physically located in one state but do most of their business with

---

[5]These downstream checks have been common in large government surveys such as the census. See, e.g., Little (1990) and Rubin (1987), where the main use of the data is to make statistical inferences on a sampled population.

[6]For example, Lancaster, Redman, and Schein (1980).

customers in another state. Measuring deliveries to the supplier, as the EIA tried to do, led to significant misstatements of energy flows.

Other examples include the use of insurance forms and patient charts to determine treatment effectiveness, where careful examination has revealed that coding of diseases and procedures varies widely with doctors, hospitals, and location and reflects influences such as the "peculiarities of the reimbursement system."[7] Highway safety researchers have noted similar problems with accident reports, where practices for identifying accident locations vary widely between localities and jurisdictions, making it difficult to relate highway features with accident characteristics.[8]

Conceptual problems are more subtle than operational problems and so have had little separate recognition in the literature, which has focused on the former. Further, the symptoms of conceptual problems are often similar to those of operational problems, particularly when the complaints are about inaccurate values. One indication of conceptual problems is that operational solutions (e.g., improved data-capture technology) do not resolve the difficulties.

The most common case of conceptual problems is the attempted use of data for purposes other than the ones they were designed for. In such cases, important limitations or caveats are often forgotten or disregarded, particularly when there is a time lag between collection and use, or when different organizations attempt to bend the data to new uses, based on evolving and changing demands for information. The successful solution of conceptual problems requires redefinition and possible expansion of the data element, rethinking the use of the data, or utilizing additional data sources to augment previously collected data.[9]

---

[7]Kolata (1994) and Gardner (1990).

[8]O'Day (1993).

[9]In the EIA energy survey validation studies, the researchers noted that data definitions in the surveys were formulated by EIA staff, and recommended testing the definitions with industry experts.

Organizational Data Problems

Operational and conceptual data problems are usually the ones addressed in efforts to "clean up" databases. However, we have observed another level of problem, which we term organizational. The symptoms of an organizational data problem at any point in time are those of operational and conceptual problems, but for organizational problems the operational and conceptual problems persist over time, even after fixes for both types of problems have been attempted.

In this case the data-quality problem is an organizational problem rather than a technological or definitional one, i.e., the organization(s) involved have not been able to act effectively to fix the problem. This can happen when data users and creators are in different parts of an organization or in completely different organizations. In this case, there may be no adjudicator who can balance data burden (for the creators) with data benefits (for the users). Even if technology can be brought to bear on data collection, the costs of the technology may fall on the creators alone. If data redefinition or refinement is needed, this may affect the data collection demands placed on the creators, especially if the users require data that the creators do not need to do their jobs.

Although the classification of data problems as organizational is not found in the literature, most researchers in the field of data quality recognize the value of communication and agreement between data creators and users to ensure data quality.[10] However, if the two groups are separated by strong inter- or intraorganizational barriers, this communication may be difficult to achieve.

Solutions to organizational problems will typically be difficult to implement. They will require agreement within or between organizations as to what data are required, what is acceptable data quality, and how costs are to be allocated and benefits shared. Such agreement is not impossible,[11] but the negotiations needed to reach such agreement first require a clear understanding of the data problem as being organizational.

---

[10]Redman (1992, 1995).
[11]Redman (1995).

Chapter Four

EXAMPLE OF LOGISTICS DATA QUALITY PROBLEMS: THE END ITEM CODE

INTRODUCTION

Data problems need to be properly classified before they can be effectively solved. The three-level framework for classifying data-quality problems grew out of our examination of specific logistics data elements in the Army with a reputation for having quality problems. The story of the End Item Code (EIC) demonstrates that the three-level framework, by forcing consideration of the true nature and root causes of data problems, is a tool the Army can use to deal with data-quality problems in general.

The EIC is a key data element in the Central Demand Data Base (CDDB), which was created in 1987 to capture the link between an end item[1] that has failed and the part required to fix it. The EIC provides an archetypal example of how Army logistics data can go bad. In particular, the EIC has

- operational data problems (it is often missing or incorrect in the CDDB records and other files in which it is supposed to appear);

- conceptual problems (it has definitional problems that can lead to incomplete or wrong answers when one is seeking weapon-system-level information); and

---

[1] For the moment, the reader should think of an "end item" as a self-contained piece of Army equipment that operates alone but that may also function as a subsystem of a larger entity (e.g., a "weapon system") fully configured for training or combat. Later in the chapter we will have more to say about the complicated relationship between "end items" and "weapon systems" in the Army.

- organizational problems (it has successfully resisted, for more than eight years, explicit, repeated, and varied attempts by the Army to eliminate its operational and conceptual data problems).

The story of the EIC code is a good place to begin if the Army wants to know what is and is not likely to work in the future in the continuing search for ways to improve the quality of logistics data.

In this chapter we give a detailed description of the EIC and how it is used, followed by descriptions of its operational and conceptual problems. We then briefly describe potential solutions for those problems, deferring discussion of organizational problems with the EIC to Chapter Five.

DESCRIPTION OF THE EIC

History and Purpose of the EIC

The EIC code and the CDDB came into being in the context of an Army decision to begin utilizing "field usage data" for centrally computing retail stockage levels rather than continuing to use engineering estimates made during the acquisition process. The Army's goal was to reduce the costs and improve the ability of retail supply systems (i.e., supply systems at the organizational, direct, and general support levels) to satisfy demands for parts in peacetime and war. The creation of the CDDB took place within the larger context of an Army effort to respond to a new DoD-level policy directing the pursuit of "weapon-system-oriented" management of secondary items.[2] The Army has continued to state its commitment to the goals of improved stockage policy and methods. The quality of EIC data, therefore, is important to the Army in its effort to improve logistics management and support.

---

[2] See Supply Management Policy Group (1985).

Example of Logistics Data Quality Problems: The End Item Code   19

Structure and Assignment of the EIC[3]

To be assigned an EIC, the equipment in question must first qualify as an "end item." The official Army document describing EICs[4] in effect defines what an "end item" is (in the Army) by specifying a set of criteria for deciding whether a piece of equipment qualifies for assignment of an EIC code: to qualify, the equipment must have an NSN; it must be made up of repair parts (supply class IX); it must itself be a member of either supply class II (clothing and individual equipment), V (ammunition and associated materiel), VII (major end items), or VIII (medical materiel); and it must be procured through certain specified appropriation and budget accounts.[5] If the equipment meets these specifications, the Army regulation governing central supply management[6] requires end-item managers at AMC's Major Subordinate Commands (MSCs) to request EIC assignment from the Army's Logistics Support Activity.

The first position of the three-character EIC identifies the national inventory control point (NICP) manager of the equipment (i.e., one of the Army's MSCs) and a broad materiel category at that MSC (e.g., "A" stands for TACOM combat vehicles; "B" for TACOM tactical vehicles, "4" for AMCCOM light weapons, etc.). The second and third positions specify, respectively, generic families (e.g., HEMTTS, mortars, etc.) and the specific end item at the NSN level (e.g., "AAB" is an M1A1 tank with a 120mm main gun, NSN 2350-01-087-1085). The Army currently has about 8,700 end items, ranging from the M1A1 tank to a sludge disposal tank (FW9).

Army Supply Bulletin 38-102 also notes that not all end items have an assigned EIC code. It states that if the EIC cannot be identified (from the Army Master Data File, which is the primary catalog reference for

---

[3]Description of the structure of the EIC code from Army Supply Bulletin 38-102 (1990) and DA Pamphlet 700-30 (1990).

[4]Army Supply Bulletin 38-102 (1990).

[5]Note that class IX items (repair parts and components) are specifically *excluded* from this list, meaning that spares and repair parts are not end items and do not qualify for EIC assignment.

[6]AR 710-1, *Centralized Inventory Management of the Army Supply System*.

EIC codes), the EIC field "is to be left blank."[7] Thus official Army guidance calls for a *blank* in the EIC field (rather than a definite entry such as "N/A") when the end item in question does not have an assigned EIC code.

The structure of the EIC is clearly built around the structure of AMC and its MSCs, in that EICs are assigned to items of equipment that are subject to the procurement processes at MSCs. This sets the stage for potential reporting problems when an end item serves as a subsystem of another end item. For example, the data display assembly and gun control computer group for the M1A1 tank each has its own EIC distinct from the EIC for the tank itself. The same is true for the infrared searchlight for the M60, and most vehicle radios. In these cases, the guidance for entry of the EIC code on supply requisitions is clear:

> The EIC shall apply to the lowest end item that the repair part is being installed on. *For example, when repairing a radio which is installed on a truck, use the EIC for the radio—not the EIC for the truck.*[8]

For maintenance uses, the guidance is less precise: DA PAM 738-750, *The Army Maintenance Management System (TAMMS),* only directs the mechanic to enter the EIC "from the AMDF."

EIC Data Flow in Supply and Maintenance

Originally, the EIC was entered on paper forms for both supply and maintenance transactions. That situation led to many of the original problems with the data element, notably its high missing and invalid rate. Spurred by these data problems, the addition of the EIC to requisitions and maintenance requests has become increasingly automated. For example, when parts are requested for a vehicle in a motor pool, the Unit-Level Logistics (ULLS) computer can use the

---

[7] Regulatory guidance for inserting EIC codes on issue requests also appears in AR 710-2, *Supply Policy Below the Wholesale Level,* DA Pamphlet 710-2-1, *Using Unit Supply System (Manual Procedures),* and DA Pamphlet 710-2-2, *Supply Support Activity System (Manual Procedures).*

[8] Army Supply Bulletin 38-102. Emphasis added.

Example of Logistics Data Quality Problems: The End Item Code 21

vehicle's administration number to automatically access the EIC of the vehicle and insert it on the requisition. Similarly, a maintenance request from a unit to its direct support unit is generated with ULLS and can also access the EIC of major end items automatically. This capability is fairly recent and has not eliminated all problems. Among other issues, units must now keep configuration information about owned equipment up to date, and the ULLS clerks may still have to select the proper end item from a subsystem list (which also must be kept up to date) if the end item being worked on is not a major piece of equipment such as a vehicle.

Figure 4.1 illustrates the data flow back to the CDDB of information about a unit's use of a part on an end item. Any issue request[9] at the ULLS level results in a requisition being sent back to the SARSS-1 (Standard Army Retail Supply System) computer that controls the Authorized Stockage List (ASL) stocks at the Direct Support level. If the unit carries the part in its Prescribed Load List (PLL) and the part is in stock, the part is issued to the requestor and the requisition is for a one-for-one replacement to the PLL stock. If the part is not carried in the PLL or is carried but is not in stock, the issue request is

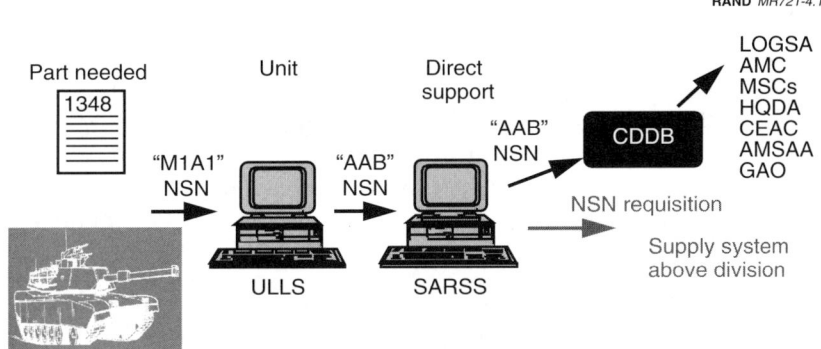

Figure 4.1—EIC Data Flow to CDDB

---

[9]Although primarily intended to capture usage of class IX repair parts, the CDDB collects images of all individual demands on supply at the organizational, direct, and general support levels in the Army, including clothing, medical, and other items. The CDDB is therefore a source of information about general retail-level supply activity in the Army.

sent to higher echelons of supply to fill the request. Note that in either case, one requisition goes back to SARSS for each demand on the PLL.

Similarly, when a unit fills out a maintenance request electronically with its ULLS, the request carries an EIC if it is available and appropriate. This is loaded into SAMS-1 when the request is processed. Parts requests at the SAMS-1 level are processed through SAMS-1 to SARSS-1. The SAMS-1 has access to the EIC on the maintenance request from the unit as well as its own internal EIC file; these are used to fill in an EIC for parts requests.

At the SARSS level (or possibly at the DS4/SAILS level if GS maintenance or other repair activities request individual parts) a copy of the request for issue will be created inside the supply computer.[10] This copy, which contains both the NSN of the requested part and the EIC (if any) of the end item associated with the request, is then passed up to DAAS for electronic transmission to the CDDB at LOGSA. Thus, the movement of CDDB data from Army installations to the CDDB is electronic and automatic. While still on the installation, the movement of CDDB data may (and often does) require the physical movement of diskettes among computers (although this data is transported with the usual supply and maintenance transactions).

Once the request copy has reached LOGSA, one more step (not shown in Figure 4.1) occurs in the flow of the EIC data element before it is finally lodged in the CDDB: an "EIC Enhancement System" (EIC-ES), created in the early 1990s to improve the quality of EIC data, is applied to the code to improve the likelihood of its having a correct, nonblank value.[11] (Some MSCs, such as TACOM, apply their own EIC imputation procedure to supplement the EIC-ES.)

The EIC is therefore affected by downstream enhancement that occurs after the data element is first entered at the retail level. Thus, the EIC flow includes multiple chances for the original EIC data ele-

---

[10]The units we visited used DS4/SAILS. However, under SARSS-O the process is functionally equivalent in that the EIC is entered below SARSS-1 and is transmitted by a separate record up to the CDDB at wholesale.

[11]Communication with Mr. Don Taylor at CALIBRE Systems, Inc. The EIC-ES is described in a System Specification Document, CALIBRE Systems, Inc. (1992).

ment to be blanked, unblanked, or changed from one EIC value to another before an analyst actually attempts to use the code to extract meaningful information. Even with increasing automation, enhancement of EIC values continues, mainly because blank and incorrect EIC values continue to show up in sizable quantities in the data from the retail level.

On the maintenance side, the SAMS-1 data on repairs is rolled up to SAMS-2, and then transmitted either electronically or by diskette (now being phased out) to LOGSA to go into the WOLF. The EICs in the WOLF are not enhanced by the EIC-ES at LOGSA.

Current Uses of the EIC/CDDB

The following are uses of the EIC that are particularly affected by data problems with the EIC.

Supply Bulletin 38-102 states that the CDDB will be used in:

(1) Determining budgets and procurements.

(2) Developing Authorized Stockage Lists (ASLs) and Prescribed Load Lists.

(3) Identifying candidate items for equipment improvements.

(4) Refining failure factors.

The CDDB is not yet utilized for all of these purposes; for example, (2) is still done largely by demand histories at each unit (largely because of the mistrust by the retail level of AMC's ability to properly set ASLs, coupled with perceived persistent quality problems with the EIC and the CDDB). However, the value of having direct access to field-level demand data is continuing to expand interest in the use of the CDDB.

The most active current use of the EIC/CDDB data is to refine failure factors in Provisioning Master Records (PMRs) for Army equipment managed by the MSCs, most notably TACOM and ATCOM. Failure factors in PMRs reflect "the expected number of failures requiring removal and replacement of a support item in a next-higher assembly or end item per 100 NHAs/EIs per year."[12] Peacetime failure factors are supposed to be based on known or estimated end-item usage in peacetime. Wartime factors are supposed to be based on known or estimated usage in wartime, including stress due to combat, accident rate, and ballistic damage. These PMR failure factors in turn are critical for a large number of Army computational procedures, processes, and models relating to stockage, including

- the Support List Allowance Card (SLAC) computations for setting spares and repair parts at the unit, DS, and GS levels of maintenance;
- the Concurrent Spare Parts (CSP) process for foreign military sales;
- the War Reserve Automated Process (WRAP);
- the Selected Essential Item Stockage for Availability Method (SESAME) stockage model; and
- the Optimum Stockage Requirements Analysis Program (OSRAP) (used to compute Contingency Support Packages and Mandatory Parts Lists).

Note that failure-factor updates could be used by the retail level to improve stockage levels in *retail* supply systems in the Army, but the data to do so can be collected only at the *wholesale* level at Army MSCs, the only organizations in a position to collect the worldwide data needed to do the computations. But to get the data, the MSCs have no choice but to rely on the *retail-level* EIC/NSN data gathered in the CDDB.

On the maintenance side, EIC information in SAMS is supposed to allow the Army to sort and aggregate maintenance actions and parts demands by end item. Missing EIC data in SAMS, however, coupled with the difficulty of doing reliable weapon-system-level analyses,

---

[12]From AMC Pamphlet 700-25 (1993).

have stymied many uses of the WOLF data for maintenance analyses by end item.[13]

Another important potential use of EIC data has arisen in conjunction with the stock-funding of depot-level reparables and the institution of the Defense Business Operations Fund (DBOF) throughout DoD in the early 1990s. Under stock funding and the DBOF, retail customers in the Army must pay for the spares and repair parts they use. This change aligns field commanders with the Army staff in their interest in tracking spending by weapon system.

Some users are trying to obtain spending information by weapon system using the EIC in the CDDB and WOLF. One example is the U.S. Army Cost and Economic Analysis Center (USACEAC). The USACEAC report on FY92 costs for Ground Combat Systems, for example, covers 31 major weapon systems. For each weapon system the report includes a rollup of the costs associated with *the distinct end items* (each with its own different EIC code) that may appear on it. For example, the M1 Abrams tank has over 30 distinct subsystems[14] that may be attached when the tank is fully configured for combat.

And at the installation level, the Tactical Unit Financial Management Information System (TUFMIS) produces a report[15] (based on EIC data fed from supply transactions) that theoretically allows commanders to monitor what they are spending to operate and support their equipment.

However, both the USACEAC and TUFMIS reports are affected by missing and incorrect EIC values in the CDDB (operational problems) and by the fact that EIC codes do not always capture all the costs associated with operating a fully configured weapon system (conceptual problems).

---

[13]This is not to say that the WOLF and the CDDB do not contain useful information, just that the quality problems with the EIC reduce their usefulness for certain critical types of analysis.

[14]These end items include such things as an M1 Hardware Adapter, a Chemical Agent Alarm Unit, a Battery Analyzer-Charger, an AN/VRC-87 Radio Set, a 50-caliber Machine Gun, and thirty-four other end items, each with its own distinct National Stock Number (NSN), Line Item Number (LIN), and EIC code.

[15]"Weapon Systems Cost Report," Production Control Number AVE 52A in TUFMIS.

OPERATIONAL PROBLEMS, CAUSES, AND ATTEMPTED FIXES

Operational data problems with the EIC are blank (i.e., presumed missing), invalid, and incorrect values. When the EIC was first introduced, the primary data-quality concern seems to have been with blank (by implication, missing) values. It seems likely that the proportion missing declined over time to a steady level, although there is little documentation about this proportion other than informal observations that the fraction of missing EICs in the CDDB has remained fairly constant in the last few years. In FY94, out of 9.62 million CDDB records, 62 percent had blank EICs (5.98 million).[16]

The operational problem of missing EICs has persisted up to the current time. For example, a sample of 697 CDDB records for five days (September 1–5) in 1994 from the 1st Infantry Division at Fort Riley, Kansas, showed 68 percent of the incoming CDDB records with blank EICs. TACOM has reported a 95 percent missing EIC rate on requisitions for a battery it manages.[17] At a meeting held at LOGSA in December 1994, representatives of TACOM and ATCOM cited missing EICs in CDDB records as one of their key data problems in computing accurate failure factors for the end items they managed.

Besides concerns with missing values, concerns with the use of invalid EICs (i.e., EICs that do not correspond to any end item) and incorrect EICs also surfaced very soon after the establishment of the EIC/CDDB system, although here again there is little quantitative information on the initial scope of the problem.[18]

Early Problems

The early problems with blank EICs stemmed from the manual input process. Originally the EIC was entered on a paper requisition (DA

---

[16]Dennis Blenman, LOGSA, personal communication.

[17]Tony Cuneo, TACOM, 1995, personal communication.

[18]USAMC MRSA (1989) says that missing, invalid, and incorrect EIC values were "severely impacting" the CDDB, but provides no quantitative statistics. Christopher (1991), for example, cites the use of the invalid code "AMY" as an EIC as a problem, but with no indication of how many requisitions were received with this code.

Form 2765 or 1348-6) or on a maintenance request (DA Form 2407 or 5504). The code either had to be known or had to be looked up in the AMDF or in the EIC Supply Bulletin. Because the EIC code was (and is) widely acknowledged to be of "no use" at the retail level,[19] there was little motivation to obtain and correctly enter EIC values. This was (and is) true, in spite of Army efforts to publicize and promote the use of the EIC code as something that can benefit the retail level.[20] Early automation simply reproduced the paper forms, in that entry of the EIC was still unassisted. Under these circumstances the only fix for blank EICs was to enforce entry of an EIC by denying the capability to obtain parts or submit maintenance requests unless EIC values were entered. Although reportedly favored by EIC users at the wholesale level, the enforcement solution was resisted by ULLS system developers at CASCOM and Development Center Lee as placing too great a data burden on the soldier at the retail level. As a result, the EIC field was left unedited: both blanks and invalid or incorrect codes were accepted.

EIC Enhancement

With data-quality editing blocked at the entry level, attention focused on checks that could be imposed downstream after the data had been entered at the retail level. One approach, used by TACOM on the CDDB data for some items, was to fill in the EIC on demands with blank EICs in the same proportions as demands where the EIC was not blank. While formally eliminating "missing" EICs, the method assumes that the missing EICs are missing at random, i.e., that any demand for a given item has the same chance of having the EIC missing, no matter what unit submitted the requisition. This is a strong assumption, particularly for items where 95 percent of the data are missing, as in the TACOM battery example cited earlier.

---

[19]There seems to be universal agreement on this point whether one is talking to wholesale-level or retail-level personnel.

[20]The 1989 MRSA Information Update on the EIC states that "it is important that users and PLL clerks understand that incorrect EIC reporting will eventually adversely impact them" and reports efforts to publicize the EIC in *PS Magazine, Army Logistician*, and other media.

The Logistics Control Activity (LCA), which was responsible for the CDDB at the time,[21] tried a different approach, in which ancillary information (from the AMDF, the CBS-X, and other Army data files) was used to impute correct EIC values in CDDB records. The EIC Enhancement System, developed by CALIBRE Systems, Inc. in the early 1990s, constructs DODAAC-to-UIC (Unit Identification Code) links and NIIN-to-EIC links. These links are used to determine what equipment is assigned to a unit submitting a demand and the parts breakdown applicable to that equipment. The EIC-ES works on about 2,500 of the Army's most important end items; it does not cover all 8,700 EICs. In some cases, a correct EIC can be imputed even if the submitted EIC is blank, invalid, or incorrect by checking what end items are owned by the unit and whether any of the unit's end items employ the requested part.[22] In other cases the enhancement code blanks out an EIC it deems wrong.

In the 1994 CDDB data from Fort Riley cited above, the enhancement process reduced the overall proportion of blank EICs by only a small amount, as shown in Table 4.1, from 68 percent in the original CDDB records to 67 percent. However, it modified the EIC in about 35 percent of the records.

The enhancement process clearly depends on timely data in the AMDF, CBS-X, and other Army files that are used by the enhancement system. This is reportedly a particular problem for the files describing the equipment assigned to units. Delays in collecting, processing and disseminating the database mean that current data are several months out of date; in addition, the detailed accuracy of the data has been questioned. The current round of downsizing, reorganization, and unit movement has added to the problems with these files.

Table 4.1 also contains data on EIC missing code rates and the effect of the enhancement system for several days in the spring of 1995 on CDDB records from Fort Riley. The missing-data rate has clearly improved (50 percent of the 1995 CDDB records had blank EIC fields), but once again enhancement had no *net* effect, as about 50 percent

---

[21]The LCA has since been incorporated into LOGSA.

[22]The logic of the imputation process is complex, particularly in cases where the requested part is not unique to a given end item.

Table 4.1

EIC Missing Rates/Enhancement Effect in Sample CDDB Data

	1994	1995
Before enhancement	68%	50%
After enhancement	67%	50%
Records modified	35%	35%

NOTE: Data from Fort Riley, Kansas, courtesy of LOGSA.

of the records were also blank after enhancement—although, as in 1994, enhancement was carried out on about 35 percent of the EIC codes.[23] We took a closer look at the enhancement process by comparing the 1995 CDDB records for Fort Riley[24] to the more detailed information available in ULLS systems located at Fort Riley. We visited both the 701st Main Support Battalion and the 101st Forward Support Battalion; senior personnel selected a small number of cases where the EIC was initially blank and cases where the enhancement process changed the EIC. They then researched those cases for us, using their closed requisition file.

In three cases where the EIC was blank and the enhancement system left it blank, the item was requested by a headquarters company and was a demand for combat boots, so the blank EIC was appropriate. In two cases, certain tools were ordered, and the enhancement system added the EIC of a wrecker. One tool did appear in the technical manual as being part of the wrecker tool kit, but the other tool was not so listed. There was no indication if the tools were in fact ordered for the wrecker. In a final set of three demands, the enhancement system had replaced the EIC (for a truck) with a blank, al-

---

[23]The fact that the 1995 CDDB records for Fort Riley had a 50 percent blank EIC rate, compared to the 68 percent rate in 1994, is most likely due to an automation change in ULLS, which is discussed next. Note that this is substantially better than the figure of 65 percent quoted earlier for the Army as a whole. The performance of the enhancement system on the EIC data for the entire Army is about the same, however, leaving 65 percent blank.

[24]We focused on Fort Riley because we had planned to do our field visits there. We believe that the findings below would be replicated elsewhere in the Army.

though the unit did have the truck and the parts were applicable (the unit provided photocopies of the relevant TM pages).

The first case bears out the procedure quoted earlier from Supply Bulletin 38-102, where it was noted that EICs are not assigned for all end items (a soldier wearing boots is not an end item) and hence some demands *should* have blank EICs. This makes the citation of missing EIC rates misleading, however, because these rates do not distinguish between missing EIC values that are "correct" (because no EIC applies) and missing EIC values that are "incorrect" (because an EIC value does apply and should appear).

The other cases show data errors introduced by the *enhancement system's* operation. Since these were a very small number of cases and were not randomly sampled from the data,[25] they cannot be used to estimate the proportion of times the enhancement system makes these particular errors. However, they do suggest that the enhancement system's operation should be formally tested against ULLS records to obtain a credible estimate of the prevalence of these problems.

Maintenance Master Data File

The key to EIC correctness clearly lies at the unit level: when a part is demanded, the end item from which it came is known. Given the persistent operational problems with EIC codes and the increasing capabilities of the computers which run ULLS, LOGSA initiated development of the Maintenance Master Data File (MMDF), which, as one of its benefits, was designed to eliminate the hand entry of the EIC code for supply and maintenance transactions.

The MMDF contains much of the static information about "major" end items[26] such as NSN, EIC, etc. When a new vehicle is added to a unit's Equipment Data File (EDF), the ULLS clerk is supposed to enter the type of equipment, its serial number and other identifying information, usage (e.g., mileage), and end items attached to it (e.g.,

---

[25]They were selected because of the action taken by the enhancement process.

[26]The MMDF includes "the reportable items and systems in AR 700-138, all ERC A and ERC P equipment (mission essential and pacing-item equipment), and all serial-number-tracked equipment" (Walker, 1994).

machine guns). In supply and maintenance transactions, the unit refers to major end items such as vehicles by administrative numbers ("admin numbers" or "bumper numbers") that are unique to the unit.[27] Once the EDF has the EIC and other information from the MMDF, however, the ULLS computer can associate the admin number to any of the other information entered in the EDF when the end item was acquired. This allows the EIC to be filled in automatically for both supply and maintenance transactions. Unfortunately, as noted previously, for lesser end items the ULLS clerk must select the subsystem from the EDF component field for the vehicle to identify the correct ("lowest level") end item requiring the part.

The software to implement this new ULLS capability was released in October 1994, and was considered to be an important step toward fixing the EIC problem once and for all, assuming that units are entering end items correctly in the EDF when acquired and that they are keeping the configuration data up to date.[28] (There is now a movement to use the ULLS for automated readiness reporting based on the configuration data, which would form a strong incentive for keeping it correct.) The use of the MMDF and ULLS automation of EIC entry from the EDF may account for the significant decrease in blank EICs (68 percent missing to 50 percent missing) in the Fort Riley sample from 1994 to 1995.

However, during our visit to Fort Riley we discovered that an important class of requisitions is being passed up from ULLS without EICs, apparently erroneously.

An Unresolved Problem: PLL Replacements

During our informal audit of enhancement performance, our hosts at Fort Riley were struck by the number of blank EIC values in the Fort Riley CDDB data (even after enhancement) for parts that they recognized and knew had been ordered for end items in their posses-

___

[27]Note that the admin number is information that does matter to the mechanic because it tells where the part must go once it has been delivered by the supply system. Maintainers thus have a natural, job-related incentive to get the admin number right.

[28]Communication with LOGSA and DCL personnel suggests that neither of these assumptions may be justified.

sion. When we checked some of these blank CDDB EICs against the original Fort Riley records, we found that they were associated with PLL replacements. PLL replacements are one-for-one, that is, when the original demand was filled from the PLL, a request is automatically generated for a replacement for the issued part for the PLL. Since the replacement is one-for-one, it should be associated with an EIC if the original demand was, otherwise the demand will not be counted against the correct EIC when it reaches the CDDB. We found that the PLL replacement request went forward from the PLL to the ASL and did eventually appear as a CDDB record, but the EIC value on the replacement request was usually blank. (The EIC enhancement system at LOGSA was able to fill in some, but not all, of those blanks with imputed EIC values.) Figure 4.2 displays where the EIC code is deleted.

We verified that one-for-one PLL replacements carry a blank EIC by deliberately generating three dummy requests at an ULLS box: one for a part in stock in the PLL, one for a part carried by the PLL, but currently out of stock, and one for a part not carried in the PLL. In each case, the part was for use on the same vehicle (a vehicle with a well-defined EIC), and in each case the ULLS clerk entered the admin number for the vehicle, so that the EIC was in principle available to the ULLS software. We then observed the data that went forward on diskette from ULLS to SARSS. The first request did not carry an EIC; the second and third requests did.

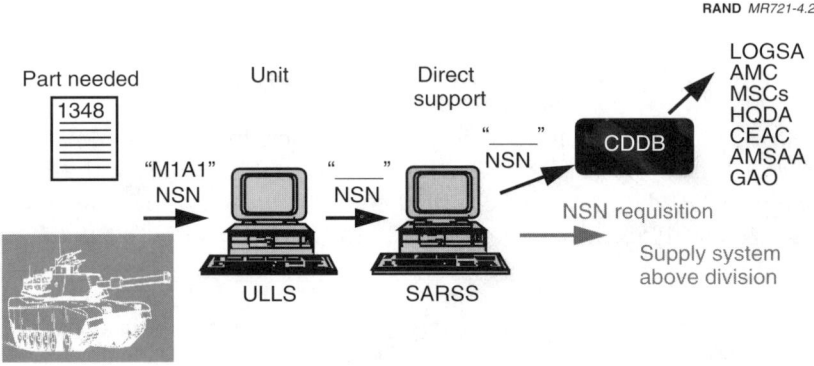

Figure 4.2—PLL Replacement Requisitions Do Not Carry an EIC

Example of Logistics Data Quality Problems: The End Item Code 33

The purpose of having the EIC on a demand is to tie the demand to an end item in the CDDB. Stock replenishments above the PLL (e.g., requisitions to replenish an ASL) cannot be tied to an EIC, because they replace stocks that satisfied individual requests bearing an EIC. Such stock replenishments are in fact correctly excluded by policy from the CDDB. However, PLL replacement demands clearly should be counted in the CDDB. The end item is unambiguous and, if it is not identified, the CDDB will miss capturing demands satisfied by the PLL. While the documentation we have examined never says explicitly whether PLL replacements should be counted or not, the wholesale system clearly wants to count each PLL replacement against the end item generating the demand. This omission is also likely to be a significant source of error in accounting for parts use at the unit level. Even more important, it could account for a majority of the remaining blank EICs in the CDDB, since the fill rate for a well-maintained PLL is claimed to be around 40–45 percent. (The TACOM battery cited earlier with a 95 percent blank EIC rate is an item routinely carried in unit PLLs, so its EIC problem could be almost entirely due to the mechanism described here.)

The extent of knowledge of this data gap is unclear. Virtually all of the personnel we talked with at the wholesale level were unaware that PLL replacements were not having EICs inserted. (The question was explicitly raised, for example, at the December 1994 LOGSA meeting, and all participants asserted that PLL replacements should and did have EICs attached.) However, personnel at Fort Riley and Fort Campbell were equally well aware that the EIC was not being attached. The same was true for ULLS systems developers at Development Center Lee. One source at DCL stated that blanked EICs on PLL replacements was a known (but low-priority) problem in ULLS (Ground) that was slated to be fixed in 1996.[29]

CONCEPTUAL PROBLEMS, CAUSES, AND ATTEMPTED FIXES

The previous section focused on operational problems with the EIC, i.e., missing, invalid, or incorrect entries. We argue in this section

---

[29]While this report was in draft, we were informed that this problem has been fixed in the System Change Proposal for ULLS-Ground distributed in late 1995.

that the EIC has conceptual problems as well: even when the operational problems have been dealt with, the EIC has definitional problems that frustrate or prevent users from getting certain kinds of parts-usage information from the CDDB.

In particular, because "end items" and "weapon systems" are not the same thing in the Army, users attempting to use EIC data to extract information about the parts usage of weapon systems face a dilemma: either they run the risk of incomplete or wrong answers, or they give up because they've learned the EIC/CDDB data can't always give them what they want. Either way, they have a data problem, but one that is caused not by missing, invalid, or incorrect entries, but rather by ambiguities and imprecision in the definition of the data element itself. This creates confusion at the "front end" when data are entered and at the "back end" when the data are used to obtain information.

An example illustrates the problem: the M1A1 tank (EIC "AAB") has thirty-odd different subsystems that may be attached when the system is fully configured for combat or training.[30] Each of these subsystems has its own EIC, different from the EIC for the tank. For example, one such subsystem, the Gun Direction Computer Group, EIC "HOU," is an end item acquired and managed by the Communications and Electronics Command (CECOM).

Now suppose one wishes to use the CDDB to assemble a report showing the total parts costs accumulated by the M1A1 tank in a given month across the Army. A natural way to do that would be to pull all the CDDB records for that month with EIC values of "AAB," compute the extended cost of the parts appearing on each record (i.e., unit cost multiplied by demand quantity), and total the result. Unfortunately, this approach will *miss* all the demands for parts that were (correctly[31]) reported against the Gun Direction Computer Group (EIC "HOU").

---

[30]USACEAC FY92 Cost Report, Vol. 2, Ground Combat Systems, August 1993. The presence of thirty-plus different possible subsystems, each with its own EIC, on one weapon system is not atypical. The USACEAC report lists from 10 to 40 subsystems or more (each with its own Line Item Number(LIN) and EIC) for each of the 23 ground-combat weapon systems covered by the report.

[31]Recall that the official reporting rule for the EIC is that the EIC will apply to the "lowest end item that the repair part is being installed on." Note also that the defini-

Of course, many of the parts demands for the Gun Direction Control Group are likely to have been (incorrectly) reported against EIC "AAB" (the tank itself), so the picture is not black and white: a CDDB query pulling all the "AAB" records will yield some, but not all, of the parts demands generated by fully configured M1 Abrams tanks. But a CDDB query pulling all CDDB records with EIC codes "AAB" and the thirty-odd *other* EIC values (for all of the tank's possible subsystems) will *overstate* parts demands for the M1A1. That is because many of the subsystems with their own EIC will apply to other weapon systems, in addition to the M1 Abrams tank. There is no easy way to get around these problems. The definition and reporting rules for the EIC code are such that there is no reliable and easy-to-describe way to use the EIC and CDDB to obtain weapon-system-level information.

The EIC's definitional problems are also reflected and compounded in the operation of the Army's supply and maintenance information systems. In supply, the "lowest-level" reporting rule is often breached because the only information attached to the request is the EIC picked up by ULLS from the unit's Equipment Data File (EDF). Unless the PLL clerk specifically determines and points to the subsystem being worked on (the EDF configuration file must also have been kept up to date), the reported EIC is likely to be the EIC for the higher-level equipment whose admin number accompanied the request. This is because ULLS will automatically refer to the EDF and will, lacking any other information or pointer to the correct, lower-level subsystem, use the admin number for the higher-level system as the pointer to the EIC. In maintenance, SAMS work orders include space for both the lower-level EIC and the weapon-system EIC, but in the transmission of closed work orders to LOGSA, only the lower-level EIC is transmitted to LOGSA. WOLF users are more often interested in the higher-level EIC, which they don't get.

Part of the problem is that in the Army the set of "weapon systems" is not well defined (and never has been).[32] Clearly, however, not every

---

tion and assignment criteria for EIC codes make the "lowest-level" rule the only rule possible if there is to be no ambiguity in the instruction about what to report.

[32]The need to define its "weapon systems" has been a recognized problem in the Army for at least 10 years; see, for example, Horn, Frank, Clark, and Olio (1989) and their discussion of the Army's response to the "secondary item weapon system man-

one of the 8,700 items with an assigned EIC would qualify as a weapon system if the Army were to settle on a definition.[33] The EIC Enhancement System mentioned in the last section, for example, only enhances EICs for 2,500 end items "of interest" (still a staggering number). The difference between weapon systems and end items becomes even more obvious when one compares the 8,700 different EICs to the roughly 900 reportable systems listed in the Army's regulation governing materiel readiness reporting[34] or the roughly 300 major systems tracked by the U.S. Army Cost and Economic Analysis Center (CEAC) for Army programming and budgeting offices.[35]

Indeed, outside the wholesale-level support infrastructure at the MSCs (which is naturally oriented toward end items because of the acquisition role and commodity orientation at the MSCs), the Army's primary interest is at the "weapon system" level (including land and aviation weapon systems, and mobility and communication systems). This applies both at the Army Staff level, where programs and budgets are assembled and Army materiel readiness is tracked, and in the retail Army, where commanders are much more likely to be concerned with the fully configured systems their troops actually use, rather than the end items defined by EICs. Indeed, as noted in the discussion of the causes of the EIC's operational problems, there is universal agreement that end item (EIC) data has no immediate, direct value or interest to the retail Army.

---

agement" (SIWSM) initiative announced by DoD in 1985 (DoD Supply Management Policy Group, *SIWSM Concept Paper*, 1985).

[33]The extremely large number of EICs partially explains why the EIC was not entered reliably when data entry was manual. Even with systems like the MMDF (which does contain the EIC codes for all EIC-assigned end items but fills codes automatically in ULLS only for the much smaller set of reportable, mission-essential, pacing, and serial-number tracked items) setting up the data and keeping it current is a major task. Compounding the problem is that some major components such as aviation jet engines do not have EICs (because they are class IX), even though sixteen different lathes each have their own EIC.

[34]Reportable on DA Forms 2406, 3266-1, 1352 for reporting Materiel Condition Status (i.e., mission-capability status). AR 700-138, *Army Logistics Readiness and Sustainability*, June 1993.

[35]CEAC uses the Operating and Support Management Information System (OSMIS)—the Army's portion of the DoD Visibility and Management of Operating and Support Costs (VAMOSC) program—to break out the operating and support costs of major Army weapon and support systems. The number of systems singled out in OSMIS provides one way of saying how many "weapon systems" there are in the Army.

Example of Logistics Data Quality Problems: The End Item Code   37

The work that CALIBRE Systems, Inc. has done for the Army with the Operating and Support Management Information System (OSMIS) is the closest that Army users have come to successfully using the EIC/CDDB data to account for parts usage by weapon system. To do that, however, CALIBRE has been forced to supplement the EIC/CDDB data with other data (e.g., files showing the weapon systems assigned to different units identified by their UIC) in order to be able to reliably prorate and assign to weapon systems the parts-demand data reported by EIC in the CDDB.

The EIC's predecessor, the Weapon System/Equipment Designator Code (WS/EDC),[36] was viewed as being "not precise enough"; the EIC, in contrast, is too precise to adequately capture the idea of what the "weapon systems" are in the Army. Ironically, by failing to be precise about what its weapon systems were in the mid-1980s (when DoD first embarked on the push to achieve "secondary item weapon-system management") the Army planted the seeds for the conceptual definitional problems with the EIC, even before the code itself was created.

In implicit recognition of the widespread interest in weapon-system-level information, the Army has taken the first steps necessary to deal with the EIC's conceptual problems: in both ULLS and SAMS, the embedded Equipment Data File for the unit is constructed to carry both the "lowest-level" EIC *and* the EIC for the weapon system/next higher assembly. What has not been done is to configure the ULLS and SAMS systems to *automatically* attach that "two-level" EIC data to supply requests and work orders, and transmit that two-level EIC upwards. That would also require reconfiguring the CDDB and WOLF databases to accept the expanded EIC data. (In the case of the WOLF, a System Change Proposal (SCP) is being worked to incorporate this feature into the new Standard Maintenance System scheduled to be deployed sometime in 1996.)

---

[36]The WS/EDC was a numeric two-digit code which only identified families of major end items, e.g., "33" referred to all combat tanks except the M1 Abrams. The EIC was intended to provide much more detailed information. USAMC Materiel Readiness Support Activity (1989).

SOLUTIONS FOR OPERATIONAL AND CONCEPTUAL PROBLEMS

In this section we discuss some potential solutions for the EIC data element's operational and conceptual problems.

Operational Solutions

One of the principles from the data-quality literature is the importance of properly capturing a data element at its source.[37] In the case of the EIC, the source is unambiguous: it is where the parts issue is done. Attempts to fill in such data downstream, such as the EIC enhancement system, require ancillary data that are not always available in a timely fashion and may have data-quality problems of their own. We speculate that the EIC enhancement system was built primarily because the wholesale system simply had no leverage to determine and control EIC data quality at the point of entry.

In large part, the MMDF/EDF solution has the potential to eliminate operational problems with the EIC. It makes use of computer capabilities to automate insertion of the EIC into issues and maintenance requests (at least for major vehicles) by keying that information to the administrative number of the vehicle. Use of the administrative number dovetails with standard practices at the unit level, and requires no new technology or procedure.

There are two drawbacks, however, to the use of the MMDF/EDF system for ensuring EIC accuracy. The first is that units *must* keep their configuration data current. This requirement is in line with the trend toward automating all of the maintenance and supply paperwork at all levels of the Army, and current plans provide incentives to the unit by using the configuration data as the basis for automated SORTS reporting. However, there is resistance to going in this direction, and problems with configuration data could affect the use of this technique to help with data entry of the EIC.[38]

---

[37]Redman (1992).

[38]In some interviews, informants have expressed strong doubts that configuration information is being kept up to date by most units.

Example of Logistics Data Quality Problems: The End Item Code   39

The second drawback is that the maintenance and associated supply recordkeeping for the vast majority of the 8,700 end items with EIC codes is not automated. This raises the question of which end items the Army really needs to track, a conceptual issue.

As noted above, the System Change Proposal for ULLS-G released in late 1995 appears to have fixed the problem with blank EICs on PLL replacement requisitions.

One final note about the expanding capabilities of ULLS software: It is possible to check the accuracy of CDDB records directly by having selected units run a query (using the ULLS query language) to download to diskette selected document fields. These diskettes could then be sent to LOGSA to compare the unit's ULLS records with its CDDB records, as we did for the sample of records from Fort Riley. This technique would allow a detailed analysis of the prevalence of the problems we discovered with the EIC enhancement process (and would have immediately identified the problem with blanked EIC codes on PLL replacement demands).

Conceptual Solutions

The conceptual problem with the EIC code arises because the Army wants to link supply and maintenance transactions to multiple levels of indenture, but the EIC code provides for only a single level. The dissatisfaction with the code stems from attempting to use it to get multilevel data and discovering that vital information is being missed.

The multiple-level aspect seems inevitable. Given their responsibilities for acquiring, fielding, and provisioning end items, the MSCs are legitimately interested in the performance of the many different subsystems that make up weapon systems. However, the Army as a whole needs comprehensive information on the support needed to operate entire systems. Even the MSCs are interested in whether common items like radios or batteries have problems in particular environments such as different weapon systems.

Based on the standard examples of multilevel EICs (e.g., radios in tanks), it seems likely that the capability for specifying *two* EICs would solve most of the current problems afflicting the EIC.[39] The capability to carry "two-level" EIC data already exists in ULLS: each end item tracked by ULLS can be associated with a "weapon-system" EIC as well. If this second, weapon-system EIC were extracted and attached to issue and maintenance requests, much of the weapon-system information that is now difficult to extract could easily be recovered. The changes to ULLS software are probably simple, but there could be a problem related to the continuing eighty-card column restriction on requisitions if an extra field were to be added.

This leads to the question of whether the Army really does need to track all 8,700 EICs. It seems likely that the assignment of EICs at most MSCs has automatically derived from the rules mentioned in the description of the EIC, and that little or no usable data is being collected on most of the items. The EIC enhancement system only worries about roughly 2,500 end items, and even that number is large. It seems much more reasonable that the Army should focus on its key combat, transportation, and communication systems and devote its resources to insuring that this data is comprehensive and of high quality. At the very most these systems should number about 500 (note that 300 are tracked by USACEAC).

On the other hand, in the longer run the Army may find that the current EIC code structure does not provide *enough* detail for the most important systems. For example, many tactical mobility vehicles are assigned EIC codes by type, even though they may be made by different manufacturers. Commercial firms are now building information systems that allow them to compare the maintenance records of similar vehicles from different manufacturers. The Army may want to eventually have that option as well. In this case the Army might want to move to attaching an identifier that uniquely identifies the particular end item undergoing repair. The MMDF structure would probably adapt to that easily with little effect on the field, but this would certainly require modifications to communications software and the central databases such as the CDDB and WOLF.

---

[39]Some have raised the issue of how to treat "weapon systems" that consist of several different vehicles, such as an air defense battery, which might require more than two EICs.

Chapter Five
ORGANIZATIONAL DATA PROBLEMS

EVIDENCE OF ORGANIZATIONAL PROBLEMS

In Chapter Three, where we defined the three-part classification scheme, we argued that the persistence over time of operational and conceptual data problems indicates that the real problem being faced is organizational. Chapter Four documents exactly such a persistent series of problems for the EIC (The appendix describes similar problems with other data elements we studied more briefly).

The persistence of operational and conceptual problems through several attempts to fix them is particularly striking in the case of the EIC. The EIC is a key data element in the CDDB, the only database in the Army for visibility of retail demands. Nevertheless, the EIC's problems have persisted for eight years, remaining largely impervious to multiple attempts at solution, including command emphasis and exhortation (for example, articles in the *Army Logistician*[1]), technical working groups, development of the EIC enhancement system, and the current initiative, the MMDF/EDF combination in ULLS and SAMS, which at present provides for automatic entry of the EIC only for vehicles and other equipment assigned an admin number). Most importantly, as noted in Chapter Four, during our interviews we found clear and compelling evidence of a significant data gap (missing EIC values on one-for-one PLL replacements) that the wholesale world was apparently largely unaware of for eight years. This in spite of the fact that the PLL problem is very likely the major

---

[1] Christopher (1991).

cause of the most troublesome problem with the EIC—missing EIC values in the CDDB.

The EIC's conceptual problems have also persisted. As we noted, there are difficulties with using the EIC to do weapon-systems-level analysis, and there is ambiguity in determining which is the "correct" EIC to report when a vehicle qualifies both as a major end item but also serves as a part of an even larger, multi–end item system (e.g., a PATRIOT battery).

The data elements discussed in this report are a subset of the elements that the retail Army provides to the wholesale Army for use in analyzing various aspects of support performance. We will argue below that persistent quality problems with these data elements arise in a fundamental way from how the Army is organized to provide logistics support and how it implements logistics information systems. By implication and extension, the entire category of Army logistics data is subject to the same organizational problems.

HYPOTHESES ABOUT CAUSES OF ORGANIZATIONAL DATA PROBLEMS

We hypothesize that there are two related organizational problems in the Army that contribute to the organizational data problems we have observed. These problems are:

- A deep divide between the Army's wholesale logistics system and the retail Army, a divide that is particularly troublesome when certain types of data must be exchanged between these two distinct parts of the Army.
- An organizational fragmentation of responsibility for specifying and implementing retail logistics information systems that amplifies the difficulties caused by the first problem.

Because our evidence is circumstantial and anecdotal, these are still hypotheses. However, their plausibility is enhanced by their consistency with the kinds of problems we see in the data elements we have studied. Below we outline this evidence and present ideas about ways to further verify the hypotheses. We then conclude with a sketch of the steps we believe are needed to solve organizational problems, which must be done if operational and conceptual data problems are ever to be finally and definitively solved.

Evidence for the Hypotheses

There are many differences between the retail Army and the Army's wholesale logistics system. The gap between the two is thrown into particular relief, however, when one considers the data that the retail level provides to the wholesale system.

One of the points repeatedly made to us about the EIC was that it is of no *direct* use to the retail level, since it is not used to manage either supply or maintenance activities at that level. (The EIC has potential usefulness for financial tracking at the retail level, but is not yet systematically used for that purpose.) The good work that has been done with the MMDF (to automatically load EIC data into Equipment Master Data Files and from that to automate EIC entry in supply and maintenance transactions) demonstrates by example that it is possible to hide the EIC completely from the retail level without adversely affecting retail supply and maintenance operations. However, the information contained in the EIC can only be created by the retail level: only at that point can an unambiguous link between a part and the higher assembly it goes on be made. (The fact that the EIC enhancement system still leaves EICs blank in 50 percent of CDDB records is evidence of the limits of downstream data imputation.)

In our interviews with retail personnel and with representatives for system development in TRADOC (CASCOM), the people we talked to insisted that soldiers should not be forced to select and enter one of 8,700 EIC codes, in addition to all of the other information that *is* necessary at the retail level to manage maintenance and supply. Personnel from the wholesale system, however, pointed out the critical importance of such data for analyzing the materiel needs of the retail level and expressed extreme frustration that the units either did not support the effort ("not enough systems discipline") or actively opposed efforts to make data entry mandatory.[2] The MMDF addition

---

[2]Data entry can be made mandatory by administrative methods or electronically, by either rejecting requisitions and work orders missing an EIC or by using entry edits to

to ULLS and SAMS was portrayed as representing the final, reluctant acceptance by the wholesale system of the futility of trying to get accurate EICs directly from data entry and the consequent necessity to automate the process.

The evidence for the wholesale/retail gap is anecdotal, but it was repeated everywhere we went. Further, it is supported by the consistent lack of error checks and feedback on the data elements we studied, data that were generated at the retail level and used at the wholesale level. In virtually every case, we were told that entry checks were not done because of the burden on the retail level to reenter the data. Conversely, it is hard to convince the retail level that a data element is critical if they never receive specific feedback from the wholesale level about errors (see the section on failure codes in the appendix). In particular, the example of the missing EIC on PLL replacement requisitions is strong evidence that the concerns of the wholesale system have never been communicated effectively to the retail level, nor apparently has common knowledge in the field found its way back up to the wholesale system.

The fragmentation of responsibility for the specification and implementation of retail logistics information systems is easier to demonstrate. The generators and users of retail-level data and the people responsible for development of retail-level information systems are quite decentralized:

- The retail units that actually generate retail-level demand and maintenance data belong to FORSCOM, USAEUR, etc. (with a smaller proportion of TRADOC, Reserve, and Guard units, the latter two of which have traditionally used different systems).

- The requirements for retail-level information systems are generated by CASCOM, which is part of TRADOC.

---

force the entry of valid EIC data before a requisition or work order goes forward. Note that the lack of a "not applicable" code for the EIC, noted in Chapter Three, and problems with the EIC enhancement system indicate that such measures would almost certainly not guarantee perfect data, and might not even substantially improve its quality, unless the checking were very specific to the equipment possessed by an individual unit.

- The design and implementation of the systems are carried out by Development Center Lee (DCL), which falls under the U.S. Army Information Systems Command.

- The organizations interested in using the data are in AMC, including its major subordinate commands and LOGSA.

- Procurement of the information systems falls under the Program Manager for Integrated Logistics Systems, who reports to the Army Acquisition Executive.

We have only anecdotal information about the effects of this fragmentation, but our understanding is that many decisions about ensuring data quality are made by the configuration committees for the retail-level data systems, where the parties either come to consensus or vote. Representatives of the wholesale system have expressed frustration that their data needs are given low priority because the wholesale system is one voice among many. And when consensus fails, the technical people at DCL are often forced to step in and make decisions. Cooperation is not impossible: the fielding of MMDF attests to that, although at its inception there was considerable uncertainty in the wholesale world about whether LOGSA or CASCOM would fund the modest MMDF effort.

Effects on Data Quality

If our two hypotheses are true, they should explain the persistence of operational and conceptual problems in data elements such as the EIC.

The connections to operational problems are clear. First, as noted by Redman (1995), data (and their quality) need to be recognized as an asset—i.e., something that has *both* benefits and costs. Redman emphasizes that this can be done only by coordinating and emphasizing such a view across the entire organization. In the Army's case, the relations between retail and wholesale level are more like those of data supplier (retail) and data customer (wholesale system), and hostile relations at that. The result is that neither side fully appreciates the EIC as an asset having benefits *and* costs. The retail system sees only the costs of collecting the data (the time required of soldiers and the effort required in systems development) but none of the benefits. The wholesale system sees only the benefits of having the data (improved information for provisioning, stockage, and maintenance) but none of the costs. The results are missing, invalid, or inaccurate data in the CDDB and WOLF.

With the MMDF, the wholesale system has begun to address the data-collection burden. The MMDF itself is a very recent development, however, and its support within the wholesale system has been shaky. Further, the automated data capabilities provided by the MMDF/EDF combination in ULLS still depend for their success on the field's willingness to update, maintain, and track configuration data at the subsystem level.

The failure to recognize data as an asset is aggravated by the organizational fragmentation in systems development, since the discussion and negotiation of data exchange and data quality is confounded with decisions about system development, rather than being addressed as a key issue in its own right. The committee forum used to make development decisions (e.g., the ULLS Configuration Control Board) is not suited to assessing data-quality benefits and negotiating the allocation of data-collection costs between two of its several constituents. The wholesale/retail gap drives the wholesale system to either develop downstream data-quality enhancement, such as the EIC enhancement system, or to simply abandon the use of a data element (e.g., failure codes) when there seems to be no way to make them work.

Our two hypotheses also explain the persistence of conceptual problems, although the explanation has more to do with fundamental differences in perspective rather than the inability to see data as an asset. In the case of the EIC, on the wholesale side there is a natural emphasis on the end-item perspective. Acquiring, provisioning, and modifying end items is what MSCs do. Even for MSC offices interested in fully configured systems, the structure of the Commodity Command Standard System (exemplified by the unavoidable end-item orientation of Provisioning Master Records) effectively forces analysis to take place by end item.

Organizational Data Problems   47

The fact that the EIC/CDDB is a creation of the wholesale world is clearly evident in the makeup of an EIC/CDDB Technical Working Group that has been meeting semiannually since 1988 to find ways to improve the accuracy of the CDDB. The original members were representatives from MRSA and LCA (now part of LOGSA), HQ AMC and each of the MSCs, SIMA,[3] LEA, and the U.S. Army Logistics Center (now CASCOM). Only the last two organizations are "nonwholesale" in nature—LEA representing the interests of the Army Staff, and CASCOM representing the interests of the retail Army. Although the latter two organizations conceivably might have tried to promote improved weapon-system capabilities for the EIC/CDDB system, there were no ways for them to do that except to go completely outside the system.

Going outside the system is exactly what the Army has done. The Army Staff, for example, working through USACEAC, has had a longstanding arrangement with CALIBRE Systems, Inc. to create the OSMIS system in order to obtain weapon-system-level information. CASCOM, for its part, has traditionally viewed its role as a defender of the field working to minimize "costs" (by minimizing the data burden) rather than expanding benefits (e.g., the capability to track costs by weapon system at the command, installation, and unit level). The EIC/CDDB system was deliberately created to accommodate the end-item perspective of the wholesale world—not the weapon-system perspective of the retail Army and the Army Staff. That fundamental difference in perspective is very much a part of the gulf between wholesale and retail, and it explains why the EIC's conceptual difficulties are the same today as they were in 1987.

We believe organizational fragmentation in systems development has also contributed to the persistence of conceptual problems. In the EIC's case, it is noteworthy that both ULLS and SAMS are equipped to carry two EIC values (one for the end item serving as a subsystem and one for the "weapon system/next higher assembly"), but the systems work has not been done to carry two-level EIC in-

---

[3]AMC's Systems Integration and Management Activity, responsible for the AMC Commodity Command Standard System (CCSS).

formation any further to where it could actually be used.[4] Again, it seems reasonable to speculate that one reason is the difficulty multiple organizations have in reaching a collective understanding of the value of such a capability, particularly when it involves the subtle difference between end items and weapon systems.

As noted above, the two causes for the organizational problem (the gulf between wholesale and retail and organizational fragmentation in systems development) are hypotheses, albeit supported by anecdotes and circumstantial evidence, and consistent with the organizational problems we have observed. We find them sufficiently plausible[5] to discuss below some ways in which they could be addressed. However, more detailed study of the hypotheses is warranted. This should take the form of a detailed investigation into the process of decisionmaking for retail logistics information systems. It should include how changes are initiated, ordered by priority, and implemented, and how problems of data exchange and data quality, in particular, are surfaced and addressed.

TWO PROPOSALS FOR FIXING ORGANIZATIONAL PROBLEMS

The key to fixing the organizational problems is the relationship between the retail level and the wholesale logistics system. The latter needs certain data that only the former can provide. They therefore must communicate and reach mutual agreement about their data needs while explicitly addressing costs and benefits. This implies first that the discussion must take place apart from technical decisions about information system implementation: decisions made on data needs and quality assurance procedures should be specifications presented to systems developers jointly by the retail level and wholesale system, not decided together with a clutter of technical and fielding decisions. It also implies that the negotiations need to

---

[4]Personnel responsible for the WOLF at LOGSA have told us that the first SCP to the new Standard Maintenance System (SMS) to be fielded in 1996 will provide the SMS with the capability to carry and pass forward two EIC values.

[5]Redman (1992) devotes considerable space to the problems caused by lack of communication between data users and data creators.

be at a high enough level to carry authority in both camps.[6] In particular, since the *long-term* benefits of improved support are critical to the field, considerations on the retail side should balance present data burden against long-term benefits, and should require the wholesale system to prove that the data requested provide benefits by performing ongoing assessment of their use. (It might be hard for the wholesale system to convincingly argue, for example, that the Army really does need to collect detailed demand data on 8,700 end items, given where MSCs in fact focus their attention.)

There are at least two possible approaches to structuring such agreement. In a *collegial* approach, joint retail-wholesale panels would review data needs, uses, benefits, and costs to determine whether data benefits justified burdens. The panels might be separated into broad "communities," e.g., aviation and land combat, with appropriate MSC representation on each panel. The retail-level representation would need to be led by FORSCOM or FORSCOM-TRADOC personnel to insure that the issues were settled at a high enough level to have force. The data-collection burden might be limited by a "data-collection budget," specified in terms of soldier time, to provide an incentive to automate collection or use other data in innovative ways. Data would also be subjected to long-term assessment by the panels, with unused data or data with consistently poor quality being revised or eliminated. These panels would then jointly propose implementation packages to the system development community.

A more intensive form of negotiation would be to form a joint retail-wholesale data process management or process improvement team with the charter to look at logistics data-quality problems in detail and with the authority to propose and implement changes.[7] A process management team would essentially have ongoing responsibility for data exchange and quality; it would "own" the data process. A process improvement team would be oriented more toward fixing specific problems as they were surfaced.

---

[6]Redman (1995, p. 99) asserts that "Due largely to the organizational politics, conflicts, and passions that surround data, only a corporation's senior executives can address many data-quality issues."

[7]Such a system of process improvement teams (PITs) is a key component of the Army's Velocity Management Initiative (Dumond, Eden, and Folkeson, 1995).

Some of our contacts have argued that the gap between retail and wholesale is too wide to allow collegial negotiation to take place and that, in fact, the two organizations are more like different companies. This suggests an alternative *commercial* approach. Since the retail system supplies the data and the wholesale system uses it, perhaps the wholesale system should *pay* the retail system for high-quality data. This approach has a number of advantages. By having to budget for data acquisition, the wholesale system could not avoid considering the costs of data. It would have incentives to make explicit tradeoffs among competing data needs, to continuously scrutinize data elements for usefulness, and to use other, perhaps cheaper data sources in innovative ways.

The wholesale system has paid for retail data in the past: prior to December 1994, in the Sample Data Collection program ATCOM paid civilian contractors at selected aviation units to scrub and validate unit data, which was then sent to ATCOM via a separate path outside regular data channels. (SDC has been used by all MSCs.) Reviving and expanding SDC is a variant of the pay-for-data approach, but it may be both too expensive and too intrusive to the host units. It would also not necessarily be able to make a transition to data collection during a deployment.

The pay-for-data proposal has a number of other significant disadvantages as well. These include the difficult challenges of setting prices, verifying data quality, and determining exactly who should get paid (FORSCOM, brigade commanders, ULLS clerks?). However, if a collegial approach is not feasible, the commercial approach has the appeal that it would force a careful appraisal of data-quality needs, provide incentives for innovation, and force at least a minimal level of negotiation between the two parties who are directly concerned with data quality and its effects.

Chapter Six
CONCLUSIONS AND RECOMMENDATIONS

OPERATIONAL AND CONCEPTUAL PROBLEMS

In this report we have discussed examples of operational and conceptual data-quality problems affecting Army logistics data. The example of the EIC (supported by the examples of the data elements treated in the appendix) suggests that a comprehensive inventory of wholesale data elements is needed, with the aim of eliminating unused elements and focusing attention on quality problems that directly affect decisions at any level of the Army. Our research suggests that many data elements will have serious operational, conceptual, and organizational data-quality problems.

Fixing some operational problems requires appropriate technology, and in the case of the EIC, using the MMDF and EDF seems to be promising. In contrast, approaches such as the EIC Enhancement System that work at the wholesale level require other sources of high-quality data that may not be available. However, the MMDF and EDF are not immune from data-quality problems; they simply use other data that presumably are more easily kept up to date to enter a correct EIC.[1]

The conceptual problem with the EIC is more difficult. For most cases (although this needs to be quantified), tracking demands and

---

[1] We have been told that not all units faithfully enter and update configuration data in their equipment files. A solution to this problem that has been discussed is to enforce configuration discipline by making it a prerequisite for the creation of required materiel condition status reports. This goes back to the exhortation/discipline approach that failed for the EIC.

repairs to both the end item and the weapon system seems to require two EICs. In the longer run, as with most conceptual problems, it might be necessary to rethink the EIC code completely. This would require a commitment to defining what a weapon system is and constructing an information flow that would track maintenance and supply transactions for such items.

ORGANIZATIONAL PROBLEMS

Data-quality problems with the EIC have persisted over the eight years the CDDB has existed, particularly the problem of blank EICs on one-for-one PLL replacements. This persistence, and the failure of previous attempts to fix the EIC, are the symptoms of an organizational data problem, i.e., the many organizations have not been able to work together effectively to solve the problem. We believe that these problems stem from deep organizational disconnects that create an environment in which input errors and definitional ambiguities are hard to avoid and difficult to eliminate. We also believe that this is the root cause of the persistence of the other problems and that it must be addressed first for any data element before most operational and conceptual problems can be solved. Problems with other data elements provided by the retail level for use at the wholesale level, particularly the failure code, share this pattern of persistence over time, as discussed in the appendix.

We have identified two organizational problems that contribute to the organizational data problems we have studied. First, there is a deep division between the retail Army and the wholesale Army, arising from fundamental differences in organization, mission, culture, personnel, and data systems. It is not clear that this gap needs to be as deep and wide as it is, but the private communications we have had from both sides indicate that it really does exist. In the case of data provided by the retail level exclusively for use by the wholesale level, treating the two groups as part of the same organization has led to logistics data not being fully comprehended by all parties as an asset. The retail level bears and complains about the collection costs, without recognizing the (deferred) benefits, while the wholesale system pays none of the collection costs but continues to defend the benefits, believing (but never being able to quantitatively demonstrate, because of data problems) that the benefits justify the costs.

Conclusions and Recommendations 53

The second organizational problem is the fragmentation of responsibility for information systems development and implementation. As noted in Chapter Five, a configuration control board involving a large number of different organizations controls the design and implementation of ULLS and SAMS, the data systems within which supply and maintenance data originate. This fragmentation aggravates the communication gap between wholesale and retail because it blurs effective two-way exchange on data needs and costs by making it part of complicated, multiorganizational communications about the many other issues surrounding data systems development and acquisition.

Instead, negotiations should occur directly between retail and wholesale about data needs and burdens. The retail level needs to understand the benefits to be derived from various data elements, and to take responsibility for the consequences to materiel support if the data are not supplied. The wholesale system needs to deal with the limits to data collection due to data burdens on the units. We have suggested some alternatives for structuring the negotiations to reach agreement on data quality, including the idea of having the wholesale system pay the retail system for data that meet quality standards. While this has significant operational obstacles that would need to be overcome, it provides incentives for new approaches to reducing data burden and increasing data quality. In this negotiation, the system development and acquisition organizations would be important advisers but would be relieved of their current role as de facto arbiters and brokers of what the systems actually do, a role that has been forced upon them by the organizational fragmentation.

The test for whether these approaches address organizational problems is whether they lead to timely, effective solutions to operational and conceptual data problems as they are discovered. The negotiations will need to move beyond data burden to a continual assessment of data quality and data usefulness, and to revisit previous decisions if quality problems arise or the usefulness of a data element does not meet expectations.

DATA QUALITY AND INFORMATION QUALITY: IMPLICATIONS FOR FORCE XXI AND VELOCITY MANAGEMENT

Force XXI characterizes information as an asset, perhaps the key asset of armed forces of the 21st century.[2] While much of the attention has been focused on tactical and strategic information in support of effective combat operations, logistics information is just as much a key asset for support operations. But to be an asset, information must be built upon data of good quality. To have effective and efficient support, therefore, the Army will need to improve data quality in all parts of its logistics information systems. The kinds of problems we have discussed in this report will need to be attacked aggressively when discovered.

Data quality also has implications for the Army's efforts to achieve velocity management (VM) in logistics. Central to the VM effort is the idea of measuring logistics processes in order to establish baselines, diagnosing areas for improvement, and evaluating achievement. Performance measurement relies on data. VM process teams may find it useful to apply the three-level framework when confronted with data problems:

- Is the problem one of missing, invalid, or inaccurate data (operational)?
- Or are data available but unusable for measuring what we want to measure (conceptual)? Have solutions been attempted but failed?
- If so, do deeper reasons exist that have caused the problems to persist over time (organizational)? What are the organizational implications for getting the problem solved?

Baseline measures for some processes may be very uncertain or even impossible to obtain, because data that have been assumed to be available may be unreliable or missing.[3] In some cases opera-

---

[2] See, for example, Grobmeier (1994) and TRADOC (1994).

[3] Pipeline segment times for the on-installation segments in the LOGSA Logistics Intelligence File (LIF) are often are not present because they are not consistently reported from the retail level.

tional problems are, in fact, conceptual problems when the required data have never been recognized as necessary and (therefore) have never been defined, much less collected.

Improvement of data may require a separate effort in VM, along with other process improvements. If the real problems are organizational, particularly with data that are primarily used at the wholesale level, improving data quality may require a systems approach across organizations and processes, rather than fixing individual data elements.

Appendix
OTHER DATA ELEMENTS

Our original list of problem data elements contained several different items in addition to the EIC. As our research progressed, we focused primarily on the EIC because it was the most representative of the data elements in terms of its importance to the wholesale logistics system. We also removed timeliness from the list because it was evident after we developed our classification of problem types that timeliness was an *example* of an operational data problem (in most cases) rather than a data element itself. We analyzed failure codes in almost as much detail as the EIC; for the rest we performed less detailed analyses, either because the problem was embedded in a larger one (the case of the Military Occupational Specialty) or because quality measures were more difficult to construct (the cases of part used and serial number).

FAILURE CODE

Description

After identification of the assembly being repaired, the next key element in maintenance management is a description of the problem encountered with the assembly and with its component parts. In theory, analysis of failures over many repairs under different conditions can allow analysts to characterize recurring problems to guide usage and future modifications. Accordingly, Army maintenance data systems include data elements that are designed to capture failure information. The data elements treated here are those collected by SAMS-1 at the FSB and MSB levels of repair.

There are two failure codes collected in SAMS-1.[1] One occurs on the task record, the other in the parts request. Both use the table of codes in DA PAM 738-750, *The Army Maintenance Management System (TAMMS)*. Currently there are slightly fewer than 200 codes. Together they cover both electronic and mechanical equipment. The distinction between the codes when they are used on the task record versus the part request is not made clear (a conceptual problem). The use of the code on the parts record seems to be directed to the specific failure of the part being replaced. Whether the use of the code on the task record should indicate a part failure if the task is to replace a part, or some broader failure to which the part contributed, is not as easy to determine. Both codes are among the data elements kept as the SAMS-1 data is rolled up to SAMS-2 after the work order is closed, and then transmitted to WOLF. There is no error checking of either of the codes against the list in PAM 738-750; in fact, entry of the failure code is not enforced by SAMS-1. Clerks or shop sergeants responsible for shop management transcribe the data from repair forms completed by mechanics.

Data Uses

The use of the failure code at the SAMS-1 site is problematic. Although the code is available in the work order detail report, a separate report, the work order status report (which a former maintenance officer characterized as much more useful) uses only the textual description of the malfunction, per the SAMS-1 manual. Further, closed work orders are purged when transferred to SAMS-2 (weekly, according to the SAMS-1 manual), and so most SAMS-1 sites would have little historical data available to them in order to aggregate and analyze failure patterns with the failure codes. Our reading of the SAMS-2 manual suggests that failure codes are little used at that level, as well. This is consistent with the practice of purging closed work orders monthly at the SAMS-2 site.

The usefulness of failure codes is therefore presumably at the wholesale level, where long-term history is maintained in the WOLF, and analysts have access to repair work orders from across the Army.

---

[1] A third failure code is collected when a serial-number-tracked part is replaced, but this is so rarely done that we have ignored it here.

However, while contacts at TACOM and ATCOM characterized failure codes as potentially very useful, they agreed with LOGSA staff that data problems with the failure codes made them "useless." LOGSA does not recommend use of the failure code for analyses, and has begun experiments with the textual descriptions to provide information on failure modes, bypassing the failure code altogether.

Data Problems

Operational problems. There are three basic operational problems with both of the failure codes that concern the wholesale system: missing codes, invalid codes, and uninformative (although valid) codes. As an example of the first two problems, Table A.1 shows the distribution of codes from task and part records for an MSB (three repair companies) and two FSBs, which together support a heavy division:

At these units there are quite different patterns of code problems, both between the failure codes on part and task records and between the two types of units. The records with missing failure codes are substantial at the MSB on both task and part records, while at the FSB most of the part records have failure codes filled in. However, it is not clear from a simple count whether or not a blank failure code is always incorrect: the example in PAM 738-750 shows blank failure codes when the task performed is an initial and final inspection.

Of the 1,752 invalid failure codes on the MSB task records in our data, 1,724 were "920," and these were primarily from only one of the three companies. We contacted the unit in question and found that they

Table A.1

Valid, Missing, and Invalid Failure Codes

Values	MSB		FSBs	
	Part Record	Task Record	Part Record	Task Record
Blank	50%	21%	3%	52%
Valid	45%	40%	95%	48%
Invalid	5%	39%	2%	0+%
Total	1,743	4,519	2,266	5,559

were using 920 to mean "No defect found." (This code and a list of five or so other common ones were displayed prominently around their shop.) When we pointed out that "799" was the code for "No defect," according to PAM 738-750, they asked (reasonably) why the code had been accepted for the many months they had been using it. When this code is not included, the number of invalid codes from the MSB drops in line with the task records and the usage at the FSB.

Error checking at data input could, in theory, eliminate the bad codes. Forcing a valid entry (once the question is settled of how to code inspections) could eliminate blank codes. However, since the data are not being input by the mechanic, finding a correct code if it is not on the form would require tracking down the responsible mechanic and getting a new, valid code. This would probably result in the extensive use of a small number of valid codes. This is already happening. In Table A.2 we describe the usage of valid codes for the units above.

Even when the failure code is valid (e.g., "broken"), at least in these data, it does not provide much information, certainly nowhere near the amount that 200 unique code values imply might be available.

A number of plausible causes exist. As we noted, the failure code information seems to be of no use to the maintenance units, either the mechanics or management, based on the coverage it is given in the SAMS manuals and on our conversations with current and former

Table A.2

Breakdown of Valid Failure Codes

Failure code	MSB		FSB	
	Part Record	Task Record	Part Record	Task Record
"Broken"	79%	6%	70%	99%
"Fails Diagnostic"	19%	53%	2%	0+%
"Worn excessively"	0+%	36%	22%	1%
"Leaking"	1%	1%	5%	0+%
"No Defect"		4%		
Other	1%	1%	1%	0+%
Total	774	1,799	2,151	2,658

maintenance personnel. The element is not required or checked, nor is any feedback given, even belatedly, when invalid codes are used.

Conceptual problems. Even if rigorous input editing could insure 100 percent valid failure codes on both the task and part records, the distribution of valid codes implies that the complete data would not carry much more information beyond "broken." This leads to the larger question of whether failure codes suffer from a deeper problem than being inconvenient to look up.

To the best of our knowledge, no systematic research has been done on good coding schemes for information like failure mode, nor has there been a systematic examination of commercial practices, although some experience with the mass transit industry indicates that failure codes are little used or are few in number.[2] We speculate that a long list of codes for a wide range of equipment is hard to use: it requires a manual easily at hand and, at least in the PAM 750-738 form, requires the user to sift through an alphabetized use of all codes to find the relevant one. A computerized list of relevant codes, selected based on the item under repair, is one solution, but this would require that the mechanic be the one who enters the information (not infeasible, but it would require drastic reorganization of the automation in the maintenance shop).

Further, for the units covered by our data, quite a bit of variability exists from unit to unit in what data are emphasized and how the SAMS-1 system is used to manage maintenance. For example, consider the incidence of blank failure codes in the three companies in the MSB shown in Table A.3. These differences are due to different procedures in each company for using SAMS-1.

Given that the codes being entered are not informative, and hence are not being used by the wholesale level, two alternative approaches present themselves, assuming that data on failure modes are important when failures are repeated and chronic.

The wholesale system could simply dispense with routine collection of failure mode information, and instead rely on detecting when the repair rate for a particular component shows a significant increase.

---

[2]Robbins and Galway (1995).

Table A.3

Percentage of Missing Failure Codes by MSB Company

	Part		Task	
	Missing	Not Missing	Missing	Not Missing
Company 1	55%	45%	1%	99%
Company 2	57%	43%	56%	44%
Company 3	5%	95%	1%	99%

At this point, targeted information gathering could occur, which could take the form of activating special response screens on selected SAMS-1 systems. This would at least serve the purpose of alerting the retail level that the specific information was being collected for a particular problem and might improve compliance.[3]

Alternatively, since the reason for using numeric codes is to facilitate computer selection of specific repair records, if the assignment of one of a large number of codes is onerous, then perhaps a LOGSA alternative of processing the textual malfunction description should be aggressively pursued. This text at least has the advantage that it is used by unit maintenance managers and is part of one of the common management reports. Our data show that for the MSB, the malfunction description was not blank in about 91 percent of the cases, while for the FSB the nonblank records comprised 97 percent of the total (and the descriptions were quite varied).

Organizational problems. Failure codes, like the EIC, have had problems that have persisted for some time, and, as with the EIC, part of the problem seems to stem from the division between the retail Army and wholesale logistics system. However, failure codes have not received the level of attention accorded to the EIC, primarily because of their limited use (although their use may be limited because of their poor quality). Since the wholesale system has had to do without accurate failure codes, it has been forced to effectively ig-

---

[3] For maximum usefulness, this information would have to be coupled with OPTEMPO information so that changes in repair frequency due to an intense exercise would be expected over the less stressful period in garrison. Note that this is a problem with current data usage as well.

nore failure information and manage without it. Given that failure mode data problems are not being aggressively addressed, perhaps the codes should be discarded and replaced with analysis of the text comments or collection of information on selected NSNs.

SERIAL NUMBERS

Serial numbers of parts and components do not seem to have received much attention by the Army maintenance community above the unit level. The perception is that the serial number data from SAMS is poor. To the best of our knowledge, the serial number information in SAMS and WOLF is not used by the retail system or by the MSCs.

There are two potential uses of serial numbers that depend on the situation when they are recorded.

- If the serial number is recorded when a component is received for repair, it could be used to link repair episodes to check for chronic unresolved problems (this could only be done by the wholesale level, since local SAMS data is purged weekly and monthly). This data could also be used to track the performance of parts made by different manufacturers.

- Alternatively, if the serial number is recorded whenever a part is installed or removed from an assembly *and* the usage could be recorded, serial number information could be used to manage time-limited components.

SAMS provides for both uses of serial number data, but the latter capability is virtually unused. In our data, out of 26,000 work orders, only 34 used the serial number records that indicate installation or removal of a serial number–tracked item. ATCOM does serial number tracking of time-limited components through a separate database that uses both paper and electronic input. Some of the neglect of serial number information may be due to the fact that the maintenance request form (Form 2407) has no defined field for the information: the directions instruct workers to record the serial numbers of tracked parts installed or removed in the "remarks" section of the form.

In contrast, Form 2407 does have a field for the serial number of the item being repaired, and examination shows that the quality of these data may be fairly good. We looked at the SAMS records in our data for repair work on those M1 parts that should be serial number tracked and found that all of them had serial numbers. Further, the types of serial numbers looked consistent for the most part between items of the same type. This suggests that further examination should be made of the serial number data.

MILITARY OCCUPATIONAL SPECIALTY (MOS)

The SAMS-1 system collects and transmits to WOLF information on the personnel who actually accomplished a repair. At the SAMS-1 level this includes the identity of the worker by use of an identification number local to the individual SAMS-1 system. Above SAMS-1 only certain characteristics of the worker are retained, primarily the training (MOS) of the worker.

LOGSA identified the MOS as an element of concern, although it did not appear to be in widespread use (unlike the EIC). Instead, it was requested only for special studies. Problems with the MOS were not specified in detail, although concern was expressed about missing MOS values.

In the SAMS data we examined, there were very few invalid codes. This is because the MOS data are automatically linked to the work order information by a worker ID number, local to the shop; if the worker ID data are kept correctly, the MOS should not be missing or invalid. The unit we used as our source of data used the man-hour accounting functions of SAMS, so they kept both worker time and worker specialty information carefully. This may not be the case in other units.

It is not clear that MOS data from SAMS are superior to data available from the Army personnel system about the skills of the personnel assigned to particular units. The SAMS records might be able to indicate whether a particular MOS is being used for appropriate tasks or if particular MOS skills are more efficient for doing certain repairs, but both of these require that task and time information be kept in a standard fashion. As we noted with the failure codes, this may not be the case.

Parts Used

SAMS-1 also maintains a list of parts used in a repair (taken from Form 2407). Actual orders for parts are placed thorough SAMS, so that "parts used" record-keeping is an integral part of the order process. We were told by an Army contractor that they had invested considerable time in trying to use parts lists derived from WOLF but had given up the effort. However, as with the MOS, the precise nature of the problem was hard to determine, except that the parts lists were "not credible."

The potential value of these data are considerable: they are one input into determining the operating and support costs of various weapon systems (the focus of the contractor's work), and they could be used to compare repair practices across units as an indicator of quality of repair. However, the data are currently little used because of their perceived problems.

We raised the issue on our visit to Fort Riley and were informed that they were consistent in ordering parts on the correct work order, and were nonplussed about the utility of doing anything else. Ordering parts on any open work order or using one work order for all orders could be done, but correct recordkeeping allows an ordered part to be matched easily to the repair job, and repair jobs cannot be closed unless all parts are received, cancelled, or transferred to another open job. However, this is only one installation; we have no evidence for practices elsewhere.

Because of the lack of a precise description of the problems, and the difficulty of generating plausible standards for data quality that were internally or externally consistent, we did not pursue the "parts used" data any further.

USAGE AND ORGANIZATIONAL LEVEL DATA

Usage is a blanket term for the "wear" put on a piece of equipment: mileage, rounds fired, on-off cycles, etc. It is a key data element in the development of new spare parts computations in which demands are related to usage and then projected, e.g., for contingency deployments. These data are also considered to be a key element in commercial fleet maintenance for warranty, monitoring quality, etc.[4]

Some usage information is available through the Army Oil Analysis Program (AOAP), although this does not cover all systems (e.g., the HMMWV is not in the program), nor are the data considered particularly good. ULLS's capability is being developed to record this information for vehicles, which means that the data could be attached conveniently to maintenance requests. However, determining the actual usage put on a component would require access to the ULLS data directly, i.e., it would require that the wholesale system keep and analyze unit-level maintenance data. Further, the quality of these data would depend critically on the emphasis placed on developing automated data entry for vehicles or on the individual unit's emphasis on quality of data input.

Organizational-level data certainly could be fed up to the wholesale level, particularly if the information required was fairly selective. However, interpretation of detailed data would require either rigorous and appropriate standardization of data recording practices, or familiarity with each unit's local maintenance procedures.

---

[4]Robbins and Galway (1995).

REFERENCES

Abell, John B., and Frederick W. Finnegan (1993), *Data and Data Processing Issues in the Estimation of Aircraft Recoverable Spares and Depot Repair,* Santa Monica, CA: RAND, R-4213-AF.

Army Field Manual No. 100-16 (1985), *Support Operations: Echelons Above Corps,* Washington, D.C.: Headquarters, Department of the Army, April 16.

Army Materiel Command Pamphlet No. 700-25 (1993), *Guide to Provisioning,* Department of the Army, Headquarters, U.S. Army Materiel Command, August 1.

Army Supply Bulletin No. 38-102 (1990), *End Item Codes,* Washington D.C: Headquarters, Department of the Army, May 1.

AT&T Quality Steering Committee (1992a), *Data Quality Foundations,* Holmdel, NJ: AT&T Bell Laboratories.

——— (1992b), *Describing Information Processes: The FIP Technique,* Holmdel, NJ: AT&T Bell Laboratories.

——— (1992c), *Improving Data Accuracy: The Data Tracking Technique,* Holmdel, NJ: AT&T Bell Labs.

Berger, Robert D., Edwin Gotwals, and Bobby Chin (1992), "Usage-Based Requirements Determination," *Army Logistician,* January–February, pp. 17–21.

Bigelow, James H., and Adele R. Palmer (1995), *Force Structure Costing System: Items in the Army Equipment Database*, Santa Monica, CA: RAND, PM-425-OSD.

Blazek, Linda W. (1993), *Quality Databases for Informed Decision Making*, Pittsburgh, PA: Alcoa Technical Center.

Bulkeley, William M. (1992), "Databases Are Plagued by Reign of Error," *Wall Street Journal*, May 26.

CALIBRE Systems, Inc. (1992), *End Item Code Enhancement System—System Specification Document*, Falls Church, VA: CALIBRE Systems, Inc.

Christopher, Linda L. (1991), "Using End Item Codes," *Army Logistician*, May–June 1991, pp. 6–8.

Department of the Army Pamphlet 700-30 (1990), *Logistics Control Activity (LCA) Information and Procedures*, Washington D.C.: Headquarters, Department of the Army, July 17.

Energy Information Administration (1983), *A Review of EIA Validation Studies*, Washington, D.C.: Energy Information Administration.

Dumond, John, Rick Eden, and John Folkeson (1995), *Velocity Management: An Approach for Improving the Responsiveness and Efficiency of Army Logistics Processes*, Santa Monica, CA: RAND, DB-126-1-A.

Fox, Christopher, Anany Levitin, and Thomas Redman (1994), "The Notion of Data and Its Quality Dimensions," *Information Processing and Management*, Vol. 30, No. 1, January, pp. 9–19.

Gardner, Elizabeth (1990), "UB-82 Forms Offer Wealth of Information, Misinformation," *Modern Healthcare*, September 24, pp. 18–29.

Grobmeier, LTC John R. (1994), "Engineering Information for Force XXI," *Army Research, Development, and Acquisition Bulletin*, September–October.

Hansen, Mark D., and Richard Y. Wang (1991), *Managing Data Quality: A Critical Issue for the Decade to Come*, Cambridge, MA: TDQM Research Program, Sloan School of Management, MIT.

Hardjono, Handrito (1993), *A Case Study of the Business Impact of Data Quality in the Airline Industry*, Cambridge, MA: TDQM Research Program, Sloan School of Management, MIT.

Horn, Will H., Donald T. Frank, Dorothy M. Clark, and John F. Olio (1989), *Secondary Item Weapon System Management: A New Way of Doing Business*, Washington, D.C.: LMI, AR711R1.

Kolata, Gina (1994), "New Frontier in Research: Mining Patient Records," *Wall Street Journal*, August 9.

Lancaster, M. A., J. M. Redman, and R. L. Schein (1980), *Data Validation Study of the Prime Suppliers Monthly Report (EIA-25)*, Washington, D.C.: U.S. Department of Energy, Energy Information Administration.

Laudon, Kenneth C. (1986), "Data Quality and Due Process in Large Interorganizational Record Systems," *Communications of the ACM*, Vol. 29, No. 1, pp. 4–11.

Levitin, Anany (undated), "Formats for Data Representation: A Taxonomy and Quality Dimensions," unpublished.

———, and Thomas Redman (1995), "Quality Dimensions of a Conceptual View," *Information Processing and Management*, Vol. 31, No. 1, January, pp. 81–88.

Little, R. J. (1990), "Editing and Imputation of Multivariate Data: Issues and New Approaches," in G. E. Liepens and V.R.R. Uppuluri (eds.), *Data Quality, Control Theory, and Pragmatics*, New York: Marcel Dekker.

Miller, Louis W., and John B. Abell (1992), *DRIVE (Distribution and Repair in Variable Environments)*, Santa Monica, CA: RAND, R-4158-AF.

Naylor, Sean D. (1995), "Digitized Force: Better, But Not Smaller," *Army Times*, October 12, p. 12.

O'Day, James (1993), *Accident Data Quality*, Washington, D.C.: Transportation Research Board, National Research Council.

Redman, Thomas C. (1992), *Data Quality*, New York: Bantam Books.

——— (1995), "Improve Data Quality for Competitive Advantage," *Sloan Management Review*, December, pp. 99–107.

Robbins, Marc, and Lionel Galway (1995), *Leveraging Information for Better Transit Maintenance*, Final Report, Project E-1, FY 1992, Innovative Maintenance Procedures for Standard Transit Buses, Washington, D.C.: Transit Cooperative Research Program, Transportation Research Board.

Rubin, D. B. (1987), *Multiple Imputation for Nonresponse in Surveys*, New York: John Wiley & Sons.

Supply Management Policy Group (1985), *Secondary Item Weapon System Management*, Concept Paper, Office of the Secretary of Defense, May.

TRADOC Pamphlet 525-5 (1994), *Force XXI Operations, A Concept for the Evolution of Full Dimensional Operations for the Strategic Army of the Early 21st Century*, Fort Monroe, VA: HQ Army Training and Doctrine Command, August 1.

USAMC Materiel Readiness Support Activity (MRSA) (1989), *End Item Code & Central Demand Data Base—Information Update*, Lexington, KY.

Walker, Ken (1994), *MMDF: Master Maintenance Data File*, LOGSA briefing.

Wang, Richard Y., Lisa M. Guarascio (1991), *Dimensions of Data Quality: Toward Quality Data by Design*, Cambridge, MA: TDQM Research Program, Sloan School of Management, MIT.

Wang, Richard Y., and Henry B. Kon (1992), *Toward Total Data Quality Management (TDQM)*, Cambridge, MA: TDQM Research Program, Sloan School of Management, MIT.

———, ———, and Stuart E. Madnick (1992), *Data Quality Requirements Analysis and Modeling*, Cambridge, MA: TDQM Research Program, Sloan School of Management, MIT.